PLATE TECTONICS AND GREAT EARTHQUAKES

PLATE TECTONICS AND GREAT EARTHQUAKES

50 Years of Earth-Shaking Events

Lynn R. Sykes

Columbia University Press

New York

Columbia University Press
Publishers Since 1893
New York   Chichester, West Sussex
cup.columbia.edu

Library of Congress Cataloging-in-Publication Data
Names: Sykes, L. R., author.
Title: Plate tectonics and great earthquakes : 50 years of earth-shaking events / Lynn R. Sykes.
Description: New York : Columbia University Press, [2019] | Includes bibliographical references and index.
Identifiers: LCCN 2018054004 (print) | LCCN 2018056567 (ebook) |
ISBN 9780231546874 (e-book) | ISBN 9780231186889 (cloth : alk. paper)
Subjects: LCSH: Earthquakes—History. | Sykes, L. R. | Seismic event location. |
Faults (Geology) | Plate tectonics. | Nuclear power plants—Earthquake effects. |
Earthquake prediction.
Classification: LCC QE534.3 (ebook) | LCC QE534.3 .S95 2019 (print) | DDC 551.22—dc23
LC record available at https://lccn.loc.gov/2018054004

Columbia University Press books are printed on permanent
and durable acid-free paper.
Printed in the United States of America

Cover design: Lisa Hamm
Cover image: © Daniel Sambraus/SCIENCE SOURCE

To my wife, Kathleen Mahoney Sykes

CONTENTS

PREFACE

For a number of years, I have wanted to write about my work on earthquakes and plate tectonics going back to my early days as a young scientist at the Lamont Geological Observatory of Columbia University and to bring it up to date with my latest work. Much progress has been made in understanding great earthquakes during the past 58 years. Timing and my preparation gave me the opportunity to be involved not only in groundbreaking discoveries in the development and testing of plate tectonics during the 1960s but also in the signing of a nuclear-test-ban treaty with the Soviet Union in 1974. Throughout the past 58 years, I have worked on verifying and obtaining a comprehensive nuclear-test-ban treaty, understanding great earthquakes, and more.

A variety of tasks pulled at me through the years: research on earthquakes in the greater New York City region, additional work on plate tectonics, long-term earthquake prediction as new information and techniques became available, as well as my work as a consultant to New York State regarding the likelihood of earthquakes near the Indian Point nuclear-power plants, located along the Hudson River not far from New York City. During my 40 years as a professor, before my retirement in 2005, I advised approximately thirty graduate students at Columbia University, raised funds for their support, and usually taught two classes each year in a dozen different areas of the earth sciences, environmental hazards, and the nuclear-arms race.

Now, at eighty-one years old, I have made time to reflect upon my life—professionally and personally. My undergraduate years at the Massachusetts Institute of Technology opened doors to me both scientifically and culturally. The Lamont Geological Observatory, where I landed as a graduate student in 1960 with a degree in geology and geophysics, had been formed only a dozen years earlier. I was in the right place at the right time with the birth of plate tectonics and the development of seismic methods to verify a comprehensive nuclear-test-ban treaty. My chosen field—seismology, the study of earthquakes—is the primary science and technology for detecting, locating, and identifying underground nuclear tests. Methods for examining earthquakes are very similar to those for nuclear explosions.

At Lamont, Bruce Heezen and Marie Tharp were charting the Mid-Atlantic Ridge; Maurice and John Ewing were sending research vessels to explore parts of the southern oceans to map sediment thicknesses and to chart extensions of the Mid-Oceanic Ridges in largely unmapped areas. A number of seismologists at Lamont were using surface waves to study the crust and upper mantle beneath continents and oceans. Earth scientists explored several deep-sea trenches. These studies, along with my friend Walter Pitman's "magic profile" of magnetic anomalies and my improved locations of earthquakes along the Mid-Oceanic Ridges, made possible the theories that the continents had shifted over geologic time and were moving today and that oceanic crust was young and growing.

Although these conclusions are universally accepted now, many viewed them with skepticism at the time. When I was an undergraduate at MIT, a professor told me that respectable young earth scientists should not work on vague and false concepts such as continental drift. Among the skeptics, Maurice "Doc" Ewing, the director of Lamont, strongly believed that the continents were fixed and that the oceanic crust was very old. In fact, before 1967 most earth scientists in North America were convinced continental drift was a fantasy. I, too, was a skeptic until 1966.

In mid-1966, after halting research on another project, I worked to obtain the sense of movement in earthquakes along great faults beneath the oceans and to demonstrate that new seafloor was being formed along Mid-Oceanic Ridges and that continental drift was a reality. Fortunately, the early part of my scientific career and the following decade coincided with the golden age of funding of the earth sciences in the United States.

My decision to specialize in studying earthquakes brought with it an awareness of the importance of a full halt to the testing of nuclear weapons. A companion book, *Silencing the Bomb: One Scientist's Quest to Halt Nuclear Testing* (Sykes 2017), emphasizes nuclear testing and arms control and highlights the desirability of a full test-ban treaty and a variety of possible steps toward the control of nuclear weapons and the prevention of nuclear war. Realizing the devastating consequences of a nuclear conflict, I made a major commitment to do whatever I could to effect the signing and ratification of a comprehensive nuclear-test-ban treaty. We in the seismological community could contribute to the identification of difficult-to-identify underground nuclear explosions and perhaps bring about an end to atomic testing and the nuclear-arms race. A number of heroes and villains stand out in the search for a verifiable halt to the testing of nuclear weapons.

I have tried to make this book accessible to a wide audience of educated people, including students, scientists, and those interested in the history of science. It is not intended as a textbook or scientific review of earthquakes, plate tectonics, or earthquake forecasting. I focus instead more on the science and policy implications of the studies I participated in directly. In that sense, this book is partly autobiographical.

Chapters 1 and 3 to 6 describe my work on the development and testing of plate tectonics during the 1960s. My personal life, education, and aspects of them that influenced my work on plate tectonics and earthquakes are described in chapters 2 and 14. Chapters 7 and 8 cover my studies of seismic gaps—that is, places along plate boundaries that have not been the sites of large earthquakes for many decades. Some of these gaps are likely to be sites of major and great shocks during the next few decades. I describe efforts to refine estimates of the times of occurrence of future large earthquakes by determining repeat times of past large shocks and how far along in time certain seismic gaps are in the cycle of the buildup of pressures (stresses) toward future large shocks. Chapter 9 covers my work in the 1980s as chair of the U.S. National Earthquake Prediction Evaluation Council. Chapter 10 describes work on great earthquakes of the past fifteen years, especially the very damaging shock of 2011 in Japan and its effect on the nuclear-power reactors at Fukushima. Chapters 11 to 13 cover my work on earthquakes within the North American plate and risks to nuclear-power reactors, including the Indian Point reactors to the north of New York City. Chapter 14 discusses my travels abroad to earthquake sites and other locations of scientific merit. Finally, chapter 15 covers ongoing work on long-term earthquake prediction and my assessment of the future prospects of such prediction.

Herein is my story.

PLATE TECTONICS AND GREAT EARTHQUAKES

1

TRANSFORM FAULTS

My Road to Seafloor Spreading, Continental Drift, and Plate Tectonics

The years 1966 to 1968 were transformative in the field of earth sciences as the plate tectonic hypothesis was being formulated and tested. This chapter describes my work on transform faults and implications for the addition of new seafloor along Mid-Oceanic Ridges, the drifting of continents, and plate tectonics.

Main Elements of Plate Tectonics

Knowledge about the global distribution of earthquakes was central to the development of plate tectonics. Most earthquakes, especially those beneath the oceans, are confined to narrow belts (figure 1.1 and plate 1). The earthquakes of shallow depths beneath the earth's surface along the crests of Mid-Oceanic Ridges stand out. A good example is shocks along the center of the elevated Mid-Atlantic Ridge midway between the Americas and Africa, along which new seafloor is being added.

Shallow events also occur along another type of plate boundary, such as the San Andreas Fault of California, where the Pacific plate slides horizontally to the northwest with respect to the North American plate. The San Andreas is an example of a transform fault. The most seismically active bands at what are now called *subduction zones* are the sites of the world's largest earthquakes and of shocks that extend from the earth's surface to

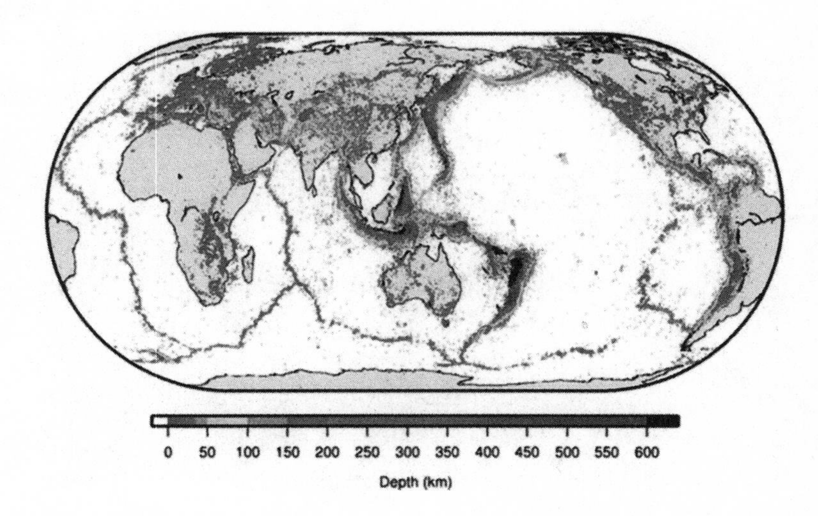

FIGURE 1.1 Global locations of earthquakes from 1960 to 2014. See plate 1, where depths of events are color-coded.

Source: From the homepage of the *Bulletin of the International Seismological Centre* website, 2014, http://www.isc.ac.uk, updated once a year.

depths as great as 430 miles (690 kilometers). Other regions, called *intraplate areas* (inside the plates themselves), are less active than the main plate boundaries.

Figure 1.2 shows the styles of displacements along the three types of plate boundaries:

1. Opening or seafloor spreading along Mid-Oceanic Ridges, as at the center of the figure, where new surface area is added to the earth by injection and cooling of liquid hot rock
2. Subduction, where one plate plunges beneath another and surface area is subtracted, as along the Tonga and Peru–Chile subduction zones, as shown schematically at the left and right sides of the figure
3. Horizontal motion along transform faults, where surface area is neither added nor subtracted

Figure 1.2 also illustrates three main layers where strength varies with depth in the earth:

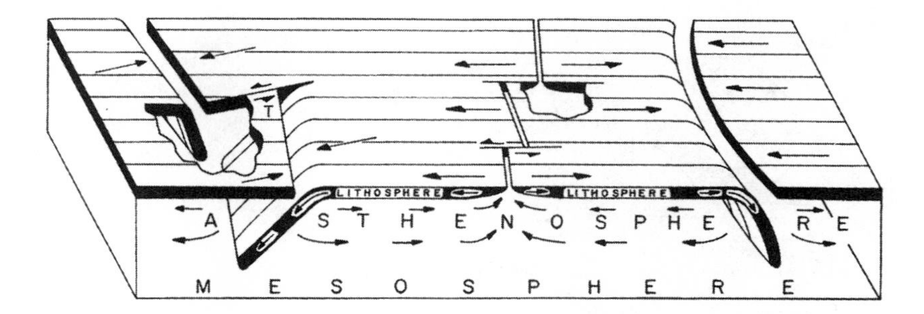

FIGURE 1.2 Schematic diagram of the main elements of plate tectonics. Arrows on the surface indicate the relative movement of adjacent plates. Three types of plate boundaries are shown—subduction, transform faulting, and seafloor spreading at ridge crests. The diagram extends from the Tonga (T) and Vanuatu subduction zones in the west (*left side*), across the East Pacific Rise in the South Pacific (*center*), to the Peru–Chile subduction zone and South America in the east (*right side*).

Source: Isacks, Oliver, and Sykes 1968.

1. *Lithosphere*: the cold, strong, outer part of the crust and uppermost mantle of the earth (the earth's plates)
2. *Asthenosphere*: the gliding layer of plate tectonics of low long-term strength where the temperature of rocks is closer to the melting point at a depth of about 60 miles (100 kilometers)
3. *Mesosphere*: somewhat stronger material at greater depth where temperature is lower than the melting point

Transform Faulting and Types of Faults

While I was in Fiji in 1965 working on deep earthquakes (chapter 4), James Dorman of Lamont Geological Observatory wrote to me about a paper, "A New Class of Faults and Their Bearing on Continental Drift," published in *Nature* that July by J. Tuzo Wilson of Toronto University. Wilson described a new class of faults that he called *transform faults*. I did not read Wilson's paper until I returned to Lamont in November 1965.

Before describing Wilson's hypothesis, I need to explain briefly about different types of faults in the earth (figure 1.3), their orientation, and the

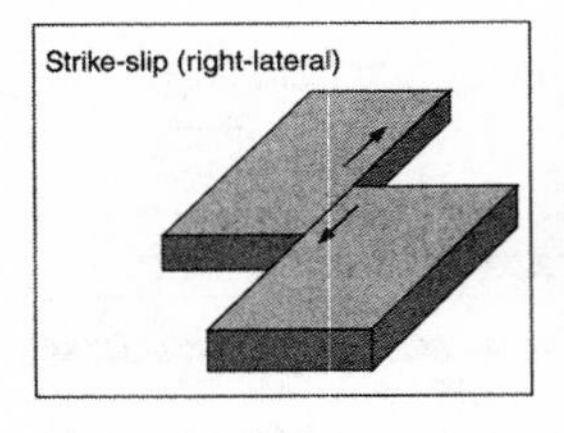

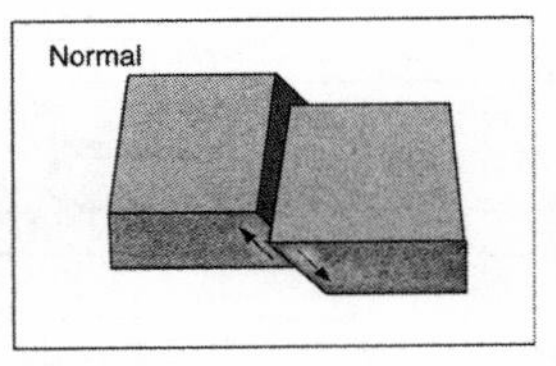

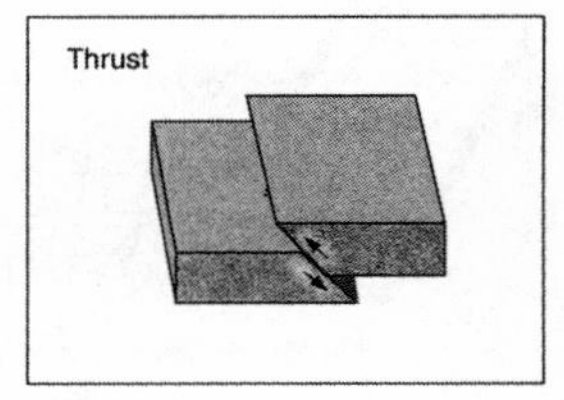

FIGURE 1.3 Main styles of faulting in the earth.

Source: Unpublished figure by the author, 2018.

mechanisms of earthquakes along them. California's San Andreas Fault is vertical and moves (slips) horizontally (upper left panel of figure 1.3), mainly during large to great earthquakes such as those in 1857 and 1906. ("Large shocks" are those greater than earthquake magnitude 7.0, whereas "great events" are larger than magnitude 7.7.) The magnitude of the San Francisco earthquake of 1906 was about 7.8.

The rocks on the two sides of the San Andreas Fault typically differ because several hundred miles (kilometers) of slip have accumulated over several million years in many past large shocks. Geologists call the direction in which a fault runs its *strike*. The orientation of the San Andreas Fault as mapped on the earth's surface is northwesterly. A vertical fault like the San Andreas, in which the direction of slip in earthquakes is horizontal, is called a *strike-slip fault*. The San Andreas is a *right-lateral* strike-slip fault, in which the Pacific plate moves northwesterly with respect to the North American plate to the east. Those terms had already been used for more than a century before Wilson wrote his hypothesis of transform faulting.

Faults in the oceans along what were formerly called *fracture zones* also involve horizontal slip along vertical faults; hence, they, too, are strike-slip faults. We now understand that those strike-slip faults are transform faults. Faults along the crests of Mid-Oceanic Ridges, which are called *normal faults* (center panel of figure 1.3), are not vertical; their fault planes instead dip at an angle of about 45 degrees from the horizontal. Their strike is parallel to ridge crests. Faults along the Mid-Oceanic Ridge system, as in figure 1.2, involve either strike-slip motion along transform faults or normal faulting along ridge crests.

Another type of fault called a *thrust fault* (upper-right panel of figure 1.3) is found at subduction zones (formerly called *island arcs*) and along major zones of crustal shortening, such as in the Himalayas. They are described more fully in chapters 4 and 5. Geologists use the term *reverse fault* when the fault's dip is steeper than that of a thrust fault. Thrust and reverse faults move in the opposite direction of normal faults.

I regard my confirmation in 1966 of Tuzo Wilson's hypothesis of transform faulting using locations of earthquakes and what are called *earthquake mechanisms* as my most important scientific contribution. My published paper "Mechanism of Earthquakes and Nature of Faulting on the Mid-Oceanic Ridges" in 1967 was one of the prime reasons I was elected to the U.S. National Academy of Sciences. The rest of this chapter describes my decisions to do that work, the methods I used, and the main results.

Mechanisms of Earthquakes and Transform Faulting

Earthquake mechanisms, which are diagrams of deformations near the source of an earthquake, can be used to determine the direction or orientation of slip during earthquakes along various types of faults. My contribution in 1966 and 1967 was to obtain mechanisms of earthquakes along faults of the Mid-Oceanic Ridge system. My work indicated that the direction of slip was what Wilson predicted for transform faults. Focal mechanisms are diagrams of fault motion at an earthquake's origin or hypocenter. (The term *hypocenter* refers to the location of an earthquake in latitude, longitude, and depth, whereas the term *epicenter* denotes only its projection onto the surface of the earth—that is, its latitude and longitude.)

Wilson had used my findings of the zigzag patterns of earthquakes along the Mid-Oceanic Ridge system in my 1963 paper, which are described in chapter 3, as key evidence for his radical new theory of transform faulting. Before Wilson's paper, the standard explanation for the present offset of two ridge segments at fracture zones is shown in the lower half of figure 1.4. Most scientists thought that the two ridge segments initially had been continuous and subsequently were displaced along a fracture zone. This model was

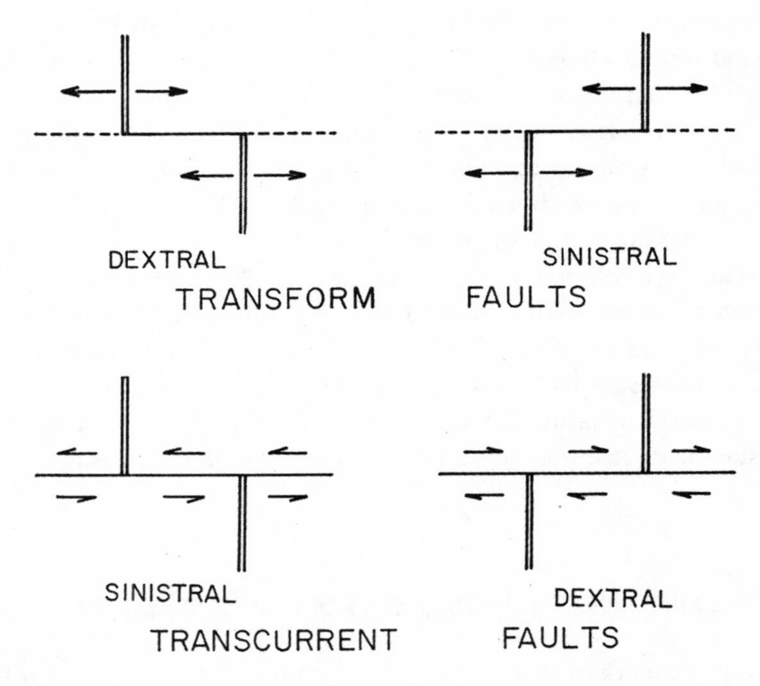

FIGURE 1.4 Sense of displacements associated with transform faults (*top*) and transcurrent faults of simple offset (*bottom*). Horizontal lines are fracture zones that intersect two ridge crests (*vertical double solid lines*). Displacement along a transform fault (*bold horizontal arrows*) is opposite that for transcurrent faults.

Source: Sykes 1967.

called *simple fault offset* or *transcurrent faulting*. If the offset of two ridge segments was the last deformation to occur, however, many earthquakes should have occurred beyond the two ridges, but they did not.

Wilson hypothesized instead that displacement along a transform fault was actually *opposite* to that for simple fault offset as shown by the bold arrows in the upper half of figure 1.4. Transform faulting was *away from* rather than *toward* each of the ridge crests. The reason, Wilson explained, was that the ridge crests weren't moving apart at all; new seafloor was instead being created along them over time as hot, liquid rock called *magma* was injected along them. Two ridge segments were born offset and remained so as the seafloor grew wider along each ridge crest. That process is called *seafloor spreading*.

Wilson proposed that only that part of a fracture zone between two ridge crests was currently seismically active and undergoing deformation (slip). Those parts of a fracture zone beyond both ridge crests are not active (dotted lines in the upper half of figure 1.4) but were created by previous seafloor spreading along those ridge crests. Horizontal motion along the active part of the transform fault today is "transformed" into horizontal spreading (opening) at each of the two ridge crests. Thus, extension, spreading, and magma injection at each ridge crest absorb displacement at the two ends of an active transform fault. It is understandable that earthquakes do not occur along a fracture zone beyond the two ridge crests in the transform model, a pattern observed first in my 1963 paper.

The natural consequences of the transform-fault hypothesis for Mid-Oceanic Ridges are that new seafloor is being generated and continental drift is occurring today.

Like many earth scientists at the time, I was initially skeptical about Wilson's hypothesis. It advocated two concepts that had many doubters: seafloor spreading and continental drift. By late 1965 and into early 1966, an idea began to germinate in my mind to prove or disprove his hypothesis using focal mechanisms of earthquakes along fracture zones. I was a good person to do this test because I *proved* rather than disproved Wilson's hypothesis, as I thought I might do originally. Returning to Lamont in November 1965, I focused on finishing my manuscript "The Seismicity and Deep Structure of Island Arcs" (Sykes 1966). Fortunately, I did that quickly and was in a position to work on transform faulting.

The "Magic Profile" of Magnetic Anomalies

My scientific life changed abruptly on a single day in the spring of 1966 when James Heirtzler, the head of the magnetics group at Lamont, called Jack Oliver to say that he and a graduate student, Walter Pitman, had some exciting new evidence about the earth's magnetic field in the southeastern Pacific. I went along with Oliver to see Pitman and Heirtzler. They proceeded to show us what became known as their "magic profile" of magnetic anomalies across the East Pacific Rise. *Magnetic anomalies* are the local variations of the magnetic field once the earth's overall field of very

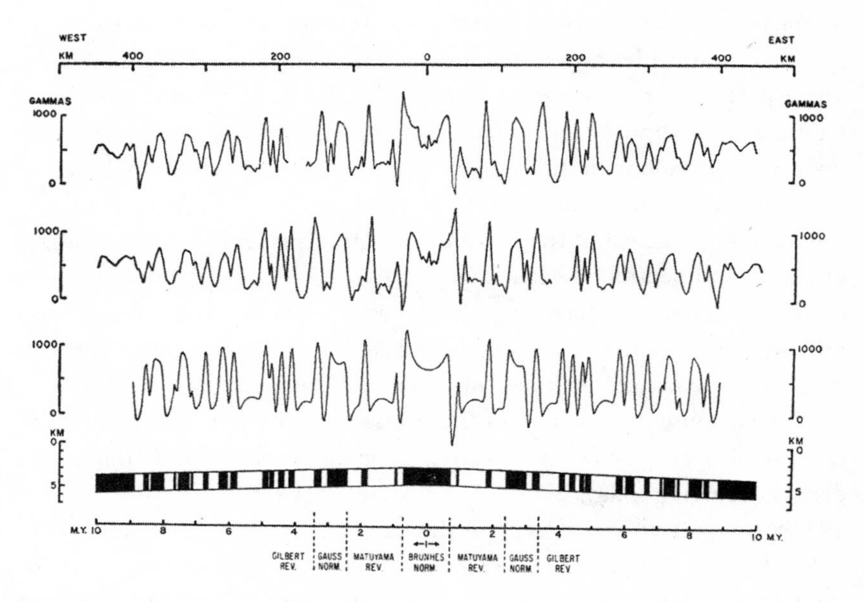

FIGURE 1.5 "Magic profile" (*center curve*) of magnetic anomalies on two sides of the crest of the East Pacific Rise in the southeastern Pacific. The original is reversed in the upper diagram so that west is plotted at the right. The lower curve is from a computer model of magnetic sources in the oceanic crust beneath the ridge crest. Black areas in the lowest subfigure represent seafloor that was magnetized in the same direction as the present earth field; the white areas are reversely magnetized. The time scale in million years, MY, extends from today at the ridge crest to 10 million years ago at each edge.

Source: Pitman and Heirtzler 1966, reprinted with permission from *Science*, the journal of the American Association for the Advancement of Science.

long wavelengths is subtracted from observations of the total magnetic field. They showed us that magnetic anomalies on either side of the East Pacific Rise were symmetrical even at the smallest detail out to distances of 300 miles (500 kilometers) from the center of that ridge. To demonstrate, they made a reverse acetate image and overlaid it on the original. The two images matched almost exactly, wiggle for wiggle, as in figure 1.5, taken from their paper (Pitman and Heirtzler 1966).

The mirror image of magnetic anomalies on the two sides of the ridge was a compelling argument to Pitman, Oliver, Neil Opdyke (a scientist who worked on what is called paleomagnetism at Lamont), and me that new material created along that ridge crest had been magnetically imprinted either positively or negatively and then had moved outward in either direction with time. That process was soon called *seafloor spreading.* Cooling seafloor was acting like a giant tape-recorder as the earth's magnetic field near the earth's surface reversed direction about every half-a-million years. Heirtzler was skeptical, saying, "I don't know what this means for continental drift." He soon changed his mind.

Fred Vine and Drummond Matthews of Cambridge University proposed the concept of magnetic imprinting of the oceanic crust in 1963. Their data came from the northwestern Indian Ocean. Their data were relatively poor and were unconvincing enough that Vine did not work on the subject again until he saw Pitman and Heirtzler's "magic profile" in 1966.

New Well-Calibrated Seismic Data and New Funding

As soon as I saw the magic magnetic profile, I knew I had to work on Wilson's transform-fault hypothesis immediately. But I needed to have high-quality seismograms from many stations around the world to read whether the motions were either upward or downward on vertical instruments. The next few paragraphs describe new seismic data that had just become available and methods I used to select the relevant seismograms for each of the earthquakes I wanted to study.

In the 1950s, the U.S. government quickly recognized seismology's potential for detecting and identifying underground nuclear explosions. The big problem was differentiating or discriminating the seismic waves produced by such explosions from the waves produced by the many earthquakes that occur every year. A panel of technical experts headed by Lloyd Berkner recommended in 1959 that the U.S. government should greatly expand funding of seismology to increase fundamental understanding of the field and to develop better instrumentation that we all hoped would eventually aid the identification of underground nuclear explosions.

Subsequent funding for the underground-explosion part of the work came from the Vela Uniform program run by the Advanced Research Project Agency of the U.S. Department of Defense. This funding transformed seismology almost instantaneously from a sleepy, poorly supported scientific backwater to a field flooded with new funds, instrumentation, professionals,

FIGURE 1.6 Stations of the World-Wide Standardized Seismograph Network (WWSSN).

Source: U.S. Geological Survey, c. 1965.

students, and excitement. Seismology and its instruments for recording earthquakes also became the main technology for detecting, locating, and identifying underground nuclear tests.

The modest seismograph network established by Lamont during the International Geophysical Year in 1957 and 1958 became the prototype for the World-Wide Standardized Seismograph Network (WWSSN). The WWSSN was installed in 1962 and 1963 with funds from the Vela Uniform program and was operated initially by the U.S. Coast and Geodetic Survey. It consisted of about 125 stations (figure 1.6) in numerous countries around the globe, each with identical Press-Ewing seismometers that measured long-period waves and Benioff seismographs for detecting short-period waves. Copies of seismograms were available to all who wished to use them in various countries and institutions. Similar data existed for new stations in Canada.

Before the advent of the new global seismic network, a seismologist wanting to study even a few earthquakes or nuclear explosions would have to write letters to many different operators of stations around the world to request copies of their seismograms. Obtaining even a minimal number of records would often take more than a year. Worse, the recordings came from a great variety of instruments, many of which were poorly calibrated or of reverse polarity (with downward motion labeled as up on records). Using the old data was often like comparing a sparse sampling of apples and oranges, some of which were partly rotten. Because reliable first motions were essential, the poor data resulted in some seismologists obtaining incorrect earthquake mechanisms.

The WWSSN solved this problem forever, providing a wealth of well-calibrated data from an expanded range of global stations and seismic sources. By essentially "viewing" the seismic waves of an event from many angles and directions around the earth, seismologists for the first time could construct reliable and detailed three-dimensional pictures of the deformation during an earthquake near its source. Lamont was one of the few places that acquired and usefully refiled microfilm copies of records from all of the WWSSN stations for the entire 20-year period of their operation.

I had come to Columbia and Lamont as a graduate student in September 1960 just as funding from the Vela program had begun. The 1960s marked an era in which the frontiers of seismology expanded rapidly. Work on global earthquakes and their locations, depths, and mechanisms were aided greatly by data from the new WWSSN instruments and funding from the U.S. Department of Defense and the National Science Foundation. The contributions from seismology to understanding earthquake mechanisms, testing the transform-fault hypothesis, and exploring plate tectonics likely would not have happened or would have occurred much more slowly without those resources. The same is true for our ability to differentiate the seismic signals of nuclear explosions from those of earthquakes.

My Work on Mechanisms of Earthquakes

An important factor for me was simply deciding to focus my work on testing Wilson's transform-fault hypothesis using both earthquake mechanisms and precise locations of earthquakes. That project had been brewing in my mind for some time and was finally triggered by my viewing of Pitman and Heirtzler's magic profile of magnetic anomalies. Testing the transform-fault hypothesis was one of the few instances in my scientific career when I immediately stopped working on everything else to concentrate on a new project. Often in science, many more interesting things come to your attention than you can work on. Deciding what to work on and when to commence an entirely new project is often difficult. Although I hadn't worked on earthquake mechanisms previously, I knew what they were and had the background to quickly familiarize myself with the relevant theory and other researchers' results.

Huge amounts of seismic data on 70-millimeter microfilm chips of the WWSSN were available in the Seismology Building at Lamont. Fortunately, the films were filed, with quite a bit of labor, by date rather than by station, which made it easy to pull out a few bundles of microfilm chips and look at many stations for a single earthquake. Lamont also had very accurate microfilm readers and printers that the Department of Defense developed especially for the WWSSN project.

I sat down with seismograms for a few earthquakes from Mid-Oceanic Ridges and read the arrival times and first motions (upward or downward) of the P waves. *P waves*, which are the fastest-propagating seismic waves, are the first to arrive at a station from either an earthquake or an explosion. The paths P waves propagate through the deep earth are illustrated in figure 3.1. At some stations, the first movements of P waves were upward on recordings of vertical motion, which indicated that those P waves left the source region of an earthquake in an outward direction—that is, as compressions. In that case, fault motion in the direction of the departing seismic wave was in an outward direction. At other stations for the same earthquake, however, the P waves were downward. Hence, they left the earthquake source as inward first motions, which seismologists call either *rarefactions* or *dilatations*. This pattern was known previously and was expected for a simple earthquake source. In contrast, P wave first movements from explosions are compressional or outward in all directions and hence are upward at recording stations.

I quickly realized that the upward or downward first motion of P waves from moderate to large earthquakes (magnitudes 5.5 to 7.0) could be ascertained more clearly from long-period records of the WWSSN. (Long-period waves vibrate about once every five to twenty seconds, whereas short-period waves vibrate about once every second.) In the lower half of figure 1.7, the long-period P wave first moves upward on line 6. Other large motions follow this first movement as time increases to the right. The much smaller earlier wiggles on that record result from earth noise, not from the earthquake. The many short upward ticks denote time marks of about two seconds put on the record every minute. The first motion of the P wave, however, is not very clear on the short-period record in the upper half of figure 1.7.

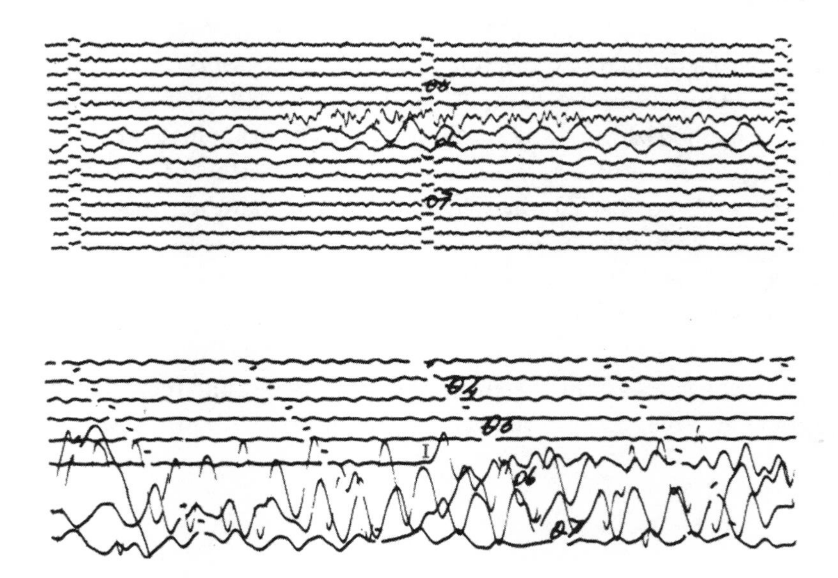

FIGURE 1.7 Comparison of first motions of P waves on a short-period vertical recording (*top*) and a long-period instrument (*bottom*) for an earthquake on March 7, 1963, at the WWSSN station at Arequipa, Peru. Time increases to the right on each line. Hour marks are numbered 04 to 07. The P wave arrived about five hours, fifty-nine minutes at the black bar. Its first motion clearly is compressional (upward) on the long-period seismogram that records vertical motion (*bottom*).

Source: Sykes 1967.

I proceeded to determine the first motion of the P wave at many stations for each earthquake using long-period recordings. Some were clearly upward (compressions), and others were clearly downward (dilatations).

It was necessary to display the first motions of P waves for a given earthquake on a two-dimensional diagram instead of on a sphere surrounding the earthquake source. One of the tools that came back to me from taking structural geology from William Brace at MIT was an *equal-area projection*. Use of it does not bias the distribution of P waves that leave in various directions from a small imaginary sphere surrounding an earthquake. Equal areas on a sphere are projected onto equal areas on that two-dimensional diagram. Finding a formula for an equal-area projection within a few days, I decided to use that method for my diagrams of the first motions of P waves.

At the University of California (UC) at Berkeley, the seismologist Perry Byerly used a method of projection he called *extended-distance projection* that was almost the inverse of equal-area projection. Although the extended-distance method badly biased the display of data from a sphere surrounding the source of an event, several seismologists had nevertheless used it. After starting work on mechanism solutions, I found, however, that William Stauder and Gilbert Bollinger at St. Louis University were using equal-area projections in their studies of seismic S (shear) waves.

I wrote a computer program to draw an equal-area projection on semi-transparent paper for each of my earthquakes. For the input latitude and longitude of each earthquake, the program calculated and then plotted on an equal-area diagram the points corresponding to where the P waves left

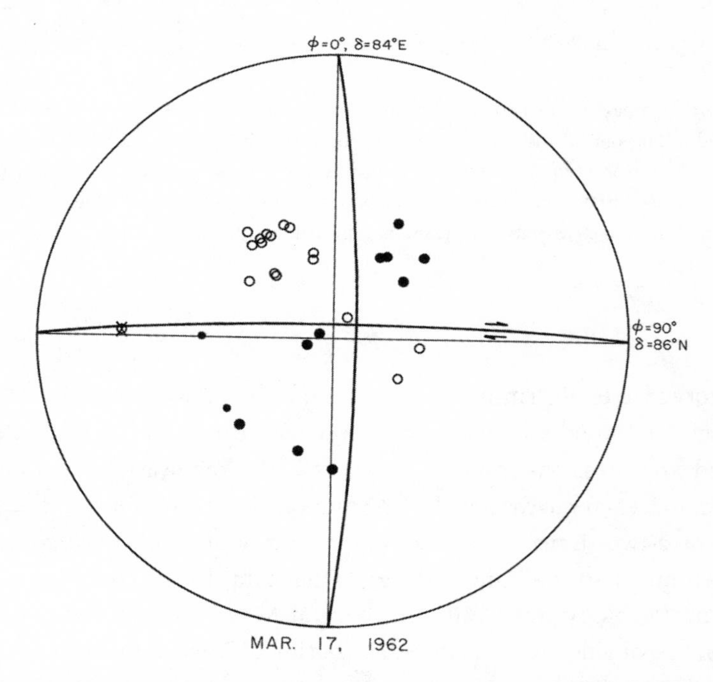

FIGURE 1.8 Mechanism of an earthquake along the Vema fracture zone of the Mid-Atlantic Ridge, trending east–west. Two nodal planes separate upward motion—that is, compressions (*solid circles*)—from downward motion—that is, dilatations (*open circles*). Opposing arrows show relative motion on the east–west plane that is parallel to the Vema fracture zone.

Source: Sykes 1967.

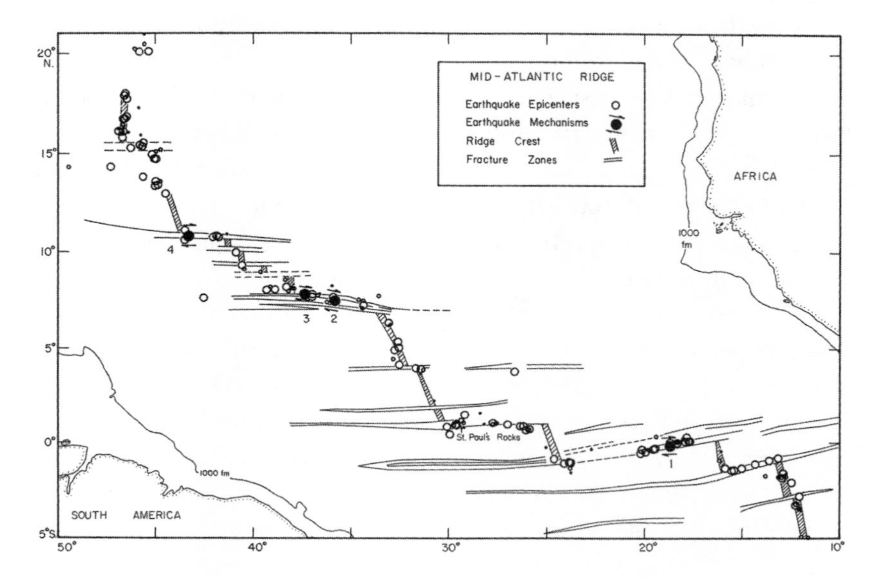

FIGURE 1.9 Relocated locations of earthquakes (*both solid and open circles*) from 1955 to 1965 and mechanisms for four strike-slip earthquakes along the Mid-Atlantic Ridge near the Equator. Opposing arrows beside each mechanism indicate sense of slip. All four solutions indicate transform faulting. Larger circles denote more precise earthquake locations.

Source: Sykes 1967.

the earthquake for all WWSSN and Canadian stations. This program saved me a huge amount of time. I then marked by hand on the diagram my own readings of the first motions—solid circles for compressions, open circles for dilatations, and X's for P waves judged to be near the transition from compressions to dilatations.

Figure 1.8 shows my mechanism diagram for an earthquake along the Vema fracture zone in the equatorial Atlantic Ocean (Shock 4 in figure 1.9). I overlaid my computer-annotated diagram of first motions on a fourteen-inch equal-area diagram and rotated the two diagrams to obtain the two best-fitting perpendicular planes, called *nodal planes*, which separated the two fields of compressions from those of dilatations. It was known at the time that first motions of P waves from earthquakes are divided into four quadrants of compressional and dilatational first motion separated by two nodal planes. One nodal plane is the *fault plane* of the earthquake, and the other is the *auxiliary plane*.

The mechanism of this earthquake involved nearly pure strike-slip faulting, with horizontal motion occurring along either of two nearly vertical nodal planes. It is not easy to determine from seismograms alone which of the two nodal planes had ruptured in that earthquake. The east–west trending nodal plane that is parallel to the Vema fracture zone, however, was the obvious choice for that earthquake's fault plane. I was nearly breathless with excitement—this was my first mechanism solution for an earthquake along a fracture zone. The sense of motion along it indicated that it was a transform fault and not a simple offset.

For this earthquake and other mechanisms described in my 1967 paper, 95 to 100 percent of the first motions fit a quadrantal pattern without any discrepancies—that is, two areas of compressions and two of dilatations. Previous studies using older data often had 25 percent or greater discrepancies. Such large uncertainties had often led to a grossly incorrect orientation of at least one of the nodal planes and an incorrect interpretation of the type of faulting.

After working out this mechanism and a few mechanisms of earthquakes along fracture zones, I knew I had made a big discovery—Wilson was *right*. It was possible to read up to 125 records for a single earthquake and make a computer plot for that earthquake in one to two days. Within a few months, I obtained fifteen earthquake mechanisms along the Mid-Oceanic Ridge system and its extension into East Africa.

In addition to strike-slip earthquakes along fracture zones, which involved transform faulting, I obtained other mechanism solutions that involved normal faulting and horizontal extension (figure 1.3). Mechanisms 6 and 7 in figure 1.10, which were located along relative straight crests of the Mid-Atlantic Ridge, are examples of normal faulting, as are several in East Africa and one along the extension of the Mid-Atlantic Ridge into the Arctic Ocean. The two nodal planes of mechanism 6 in figure 1.11 are not vertical, as in the strike-slip shock in figure 1.8. The normal-faulting mechanisms were in accord with extension and growth of new seafloor along those ridge crests.

The mechanism of one earthquake along what was called the "Macquarie Ridge" was entirely different. It involved thrust faulting associated with horizontal compression, similar to deformation farther north in New Zealand and the Tonga-Kermadec island arc. Hence, that "ridge" was not spreading as the ridge was spreading along the Mid-Oceanic Ridge system.

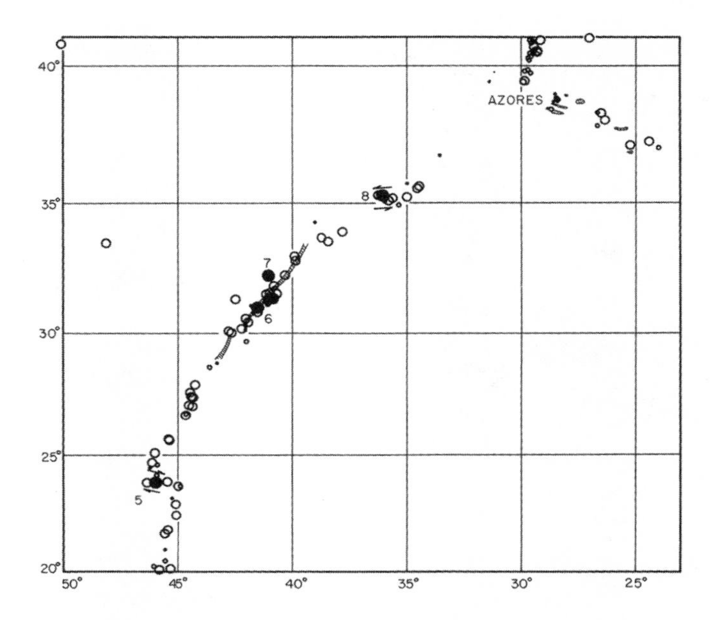

FIGURE 1.10 Relocated earthquakes (*both solid and open circles*) along the northern Mid-Atlantic Ridge from 1955 to 1965. Two strike-slip earthquakes (events 5 and 8) and two normal-faulting events (6 and 7) are shown. Larger circles denote more precise earthquake locations.

Source: Sykes 1967.

It had been named a ridge merely because it was a region of high topography on the seafloor.

The strike-slip mechanisms for events 5 and 8 shown in figure 1.10 and their associated zigzag patterns of relocated earthquakes indicated other major transform faults. The one at 23°N, which offsets the Mid-Atlantic Ridge about 125 miles (200 kilometers), had not been identified previously. A research ship sent to that site later confirmed its existence. A fracture zone at 35°N, the site of mechanism 8, was already known.

I also obtained a strike-slip mechanism for a large earthquake in northern Iceland (not shown) and identified it as a transform fault. I spent several weeks in August 1968 working in the interior of Iceland with the graduate students Peter Ward and John Kelleher of Lamont. We recorded small earthquakes as a way to understand earth deformation within Iceland's central segment of the Mid-Atlantic Ridge. The interior of Iceland is a land

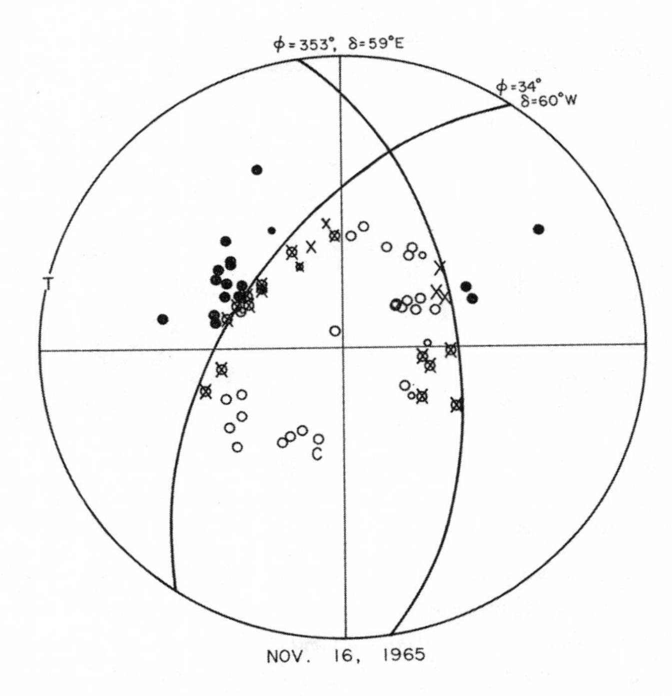

FIGURE. 1.11 Mechanism of event 6 in figure 1.10. The two nodal planes are not vertical, as in strike-slip solutions, but indicate normal faulting and extension of the oceanic crust along the ridge crest. The horizontal axis of extension, the tensional axis of the mechanism, is labeled T near the left-center edge; the axis of compression (C) is shown just below the center of the diagram.

Source: Sykes 1967.

of glaciers and very young volcanic rocks. Driving over young lava flows was slow and very bumpy. On our return to the north coast, we stopped for the night at a cabin for hikers in the interior, where it snowed in August.

In summary, my results showed that earthquakes along the seismically active parts of fracture zones involved horizontal motion along strike-slip faults in the direction predicted by Wilson for transform faulting and opposite to the direction predicted for simple offset of ridge crests. Shocks along ridge crests involved normal faulting as the two sides of the ridge spread apart in extension. I found a notable absence of earthquakes along the older parts of fracture zones beyond ridge crests. My data strongly supported the proposed processes of seafloor spreading, transform faulting, and continental drift.

Word of my study soon spread at Lamont in late summer 1966. Professor Paul Gast, a Lamont geochemist, asked me to present my work at a symposium at the NASA Goddard Space Center in New York City on November 11, 1966. My study, Pitman and Heirtzler's study on magnetic anomalies, and Opdyke's work on magnetic reversals in the sediments of deep-sea cores made a great impression on the invited audience, which included many who would become major contributors to the plate tectonics–continental drift revolution: Sir Edward Bullard, Xavier Le Pichon, Dan McKenzie, William Menard, and a number of paleomagnetists. Although the meeting was not intended to be a definitive conference on earth mobility, it was the major turning point for those concepts in the United States. Frank Press, who became the science adviser to President Jimmy Carter and who later served as president of the National Academy of Sciences, told me several years later that he regarded my paper as the first that convinced him of the reality of continental drift.

My results on earthquake mechanisms needed to be published soon. One of the best decisions in my professional life was to quickly write and submit my paper "Mechanism of Earthquakes and Nature of Faulting on the Mid-Oceanic Ridges" for publication on November 28, 1966. Fortunately, it went through the review process quickly and was published on April 15, 1967, the day before my thirtieth birthday.

Several people were anxious to have the talks presented at the Goddard meeting published as one volume. The geophysicist Robert Phinney of Princeton, a good friend of mine, agreed to get a volume published quickly by Princeton University Press. He said the papers would not have to be reviewed, which turned out not to be the case. The publisher, in fact, waited for all contributions to be submitted, had them reviewed, and then sought permissions to reproduce figures. The volume, although beautifully printed in color, was not published until 1968. By then, the entire plate tectonic revolution had occurred.

Fortunately, I also submitted to Phinney's volume a second-generation paper with additional earthquake mechanisms from the Gulf of California and the Gulf of Aden. Collections of papers from either a conference or a significant earthquake are often delayed for years, and university researchers, unlike many government employees, are penalized by not having their research grants renewed if their publications are delayed. Someone else could well have published the key and most-cited paper on mechanisms of earthquakes and transform faulting if my work had not been published quickly.

Soon after the Goddard meeting, I wanted to present my results at a national meeting. The annual meeting of the American Geophysical Union (AGU) would not be held until the spring of 1967. Getting on the program for the annual meeting of the Geological Society of America (GSA) in November 1966 was difficult. GSA, as it still is today, is quite rigid about July deadlines for submittal of abstracts, which had long passed for me. I phoned Allan Cox of Stanford University, who knew about my results from the Goddard meeting. He said, "Call up the chair of the GSA program committee, and tell them I said to put you on the program." The chair said, "You've missed the deadline, but if Allan Cox thinks it's important, we'll do it."

Fred Vine, who had attended the Goddard meeting in 1966, was invited to give a major talk on magnetic imprinting of the seafloor at that GSA meeting. He rushed to analyze magnetic data from many ridges. He published his work as the lead article in *Science* in late 1966. Vine was aware of my work. In his GSA presentation, he plugged my talk, which was included in an added session on the last day of the meeting on Saturday morning. Many people showed up for my talk, which was held in the bar of the San Francisco Hilton, which had been hastily adapted for a scientific session. I think about my talk at the bar every time I pass the San Francisco Hilton.

After my talk, David Griggs, a professor of geophysics at UCLA and a former secretary of the U.S. Air Force, came up to me with a great big smile, slapped me on the back, and said, "That's just great!" His comment certainly made my day (and in fact my decade). Griggs went back to UCLA and told his graduate students that I had proved Wilson's transform-fault hypothesis. He brought many of them to the AGU meeting in Washington, D.C., in April 1967. In 1939, he had published a major paper on mountain building and mantle convection. But when he had presented his work at a meeting of the National Academy of Sciences, Andrew Lawson of Berkeley and Barry Willis of Stanford, who rarely agreed on many things, said the paper was something akin to rubbish, so Griggs rightly now felt vindicated by my paper and the development of plate tectonics.

2

CHILDHOOD, HIGH SCHOOL, MIT, AND COLUMBIA UNIVERSITY

My parents, Margaret Woodburn and Lloyd Sykes, married in July 1936 and spent their honeymoon at Lake Lynn, West Virginia. I was born nine months later, on April 16, 1937, in Belleview, Pennsylvania, a suburb of Pittsburgh. They had decided that whatever my gender, my name would be "Lynn." My mother, born in 1910, had grown up in Belleview. She met my father, Lloyd Ascutney Sykes, while he was working in Pittsburgh forecasting weather. A year older, he grew up in the small Vermont town of West Windsor (Brownsville), Vermont. His middle name comes from Mt. Ascutney, which looms above the village. I am the eleventh generation of Sykeses in the United States. Richard Sikes, the first to arrive, was born in Yorkshire, England, in 1616 and immigrated to the Boston Bay colony in 1637. He married Phebe, and their first child was born in Roxbury. They moved to Springfield, Massachusetts, where he was granted land in 1642. Three generations later, the family changed its name to "Sykes."

In the 1920s in Brownsville, Vermont, my paternal grandfather, Ernest Sykes, who was born in 1885, operated a sawmill that he had inherited from his parents. Ernest married Louise Laber in 1907. They were very poor. Their five children helped at the sawmill, sorting wooden shingles and carrying away scrap wood for heating. My father considered an orange a special Christmas present.

My father was sent to work on his maternal grandparent's farm. After completing early-morning chores, he had to walk miles, frequently in the snow, to school. Upon graduation he was awarded a $100 scholarship to the

University of Vermont, but because he was unable to raise the remainder of his college expenses, he stayed to help his aging grandparents care for their farm for an additional year before joining the U.S. Weather Bureau in Burlington, Vermont, as a junior observer in 1928. The following year he transferred to the weather center at the Cleveland airport and was one of the first observers to brief pilots on icy conditions, turbulence, and other likely meteorological conditions en route. In 1933, he became the chief meteorologist at the newly established Allegheny airport for Pittsburgh. Active in providing local weather information to the public, he argued for an increase in weather observations in an effort to forecast the approach of major storms. His decision to alert the media after predicting an advancing cold front with plunging temperatures for the Pittsburgh area over one Thanksgiving holiday may have saved lives. Because it was warm at the start of the holiday, many vacationers in the nearby mountains had no winter clothing with them. In some ways, my father's effort was a harbinger of my own work on earthquake prediction.

My father was fortunate that he was employed throughout the Great Depression. Working long hours over maps and charts brought on eyestrain and headaches. In 1939, he returned to the U.S. Weather Bureau at the old Hoover Airport in Arlington, Virginia, near Washington, D.C., which became the site of the Pentagon in World War II. Without a bachelor's degree, he was unable to advance in the Weather Bureau, so he transferred to the Air Traffic Control Center at Washington National (now Reagan) Airport. When I was a child, he took me many times to his office on the third floor of the airport, which I found fascinating.

During the Cuban Missile Crisis of 1962, my father was responsible for keeping civilian air traffic separated from a vast armada of U.S. military planes. He told me that at the time few realized how frighteningly close the country came to all-out war. In 1964, he retired after 30 years of federal service, an excellent decision because many traffic controllers died early from heart attacks and strokes. He was proud to say he never "lost an airplane." He went on to do carpentry and remodeling for homeowners, which he enjoyed greatly.

My mother attended the University of Akron in Ohio for two years and received a teaching certificate in 1930. She left school and returned to Pittsburgh as the sole support for her parents after her father lost everything during the Great Depression. He had owned a picture-framing and print store, and his partner cheated him out of his remaining assets. My mother, who learned to play the piano as a child, gave lessons at home when we moved to Virginia. She was an accomplished organist and pianist. As an avid gardener, she was a member of the garden club of Purcellville, Virginia, and became an honorary lifetime member of the Garden Clubs of Virginia.

When my parents were married in 1936, my mother's parents moved in with them. My maternal grandmother's family, the Weitzels, had emigrated from Germany, and my maternal grandfather's family from Northern Ireland. My first memory at two years old is of my grandfather Frank Woodburn walking me to a corner store. Born in 1865 in southwestern Pennsylvania, he married Emma Weitzel, who was considerably younger than he, after his first wife died. Like me, my mother was an only child. In 1939, we all moved to Twentieth Street South, Arlington, Virginia. I remember listening to the radio with them on a Sunday afternoon in 1941 when the news that Japan had bombed Pearl Harbor was broadcast. They told me in no uncertain terms to be quiet during the terrible news.

I remember traveling by train with my father soon after the start of World War II from Washington, D.C., to Windsor, Vermont. My father wanted to visit his younger brother, Lester, before Lester was sent to the Southwest Pacific, where he was to take part in years of fierce fighting from Guadalcanal to New Guinea. We had a sleeping compartment on the Pennsylvania Railroad as far as New Haven. My father woke me to see Manhattan as we crossed the Hell Gate Bridge. It was my first view of the city where I would live many years later.

My father was deferred from military service because of his poor eyesight and age as well as his critical job in air-traffic control. He worked in Flushing, Queens, one of the boroughs of New York City, for several months during the war, where he taught airmen short courses in meteorology so that, as he said, they could learn "how to avoid being killed by the weather." My mother and I remained in Virginia, but we visited him twice. I remember being amazed to see subways, elevated trains, lights, tall buildings, and automats—things Washington lacked.

As I remember her, my maternal grandmother, Emma Woodburn, was always sick in bed. We took long trips by bus to visit her when she had to be moved into a nursing home. She died of cancer in 1946. Then my grandfather Frank was hospitalized in Washington, D.C. My mother begged her two half-sisters over the phone to help with his medical expenses. He died in 1948, and we drove to Washington, Pennsylvania, for his funeral and burial. My father took me to see him before they closed the casket. I missed him. My grandfather, who was a staunch Republican, had told me during walks that he hated Franklin Roosevelt, despite Roosevelt's introduction of Social Security, which could have rescued my grandparents if it had been available in 1929.

My paternal grandfather, Ernest Sykes, claimed to be a Jefferson–Jackson Democrat and said he was one of only two Democrats in Brownsville. He always wanted to be back in Brownsville for town meetings during March, which was known as "mud month." On his way south to Florida after his wife died, he stopped in Washington to see us and then to visit the U.S. senators from Vermont to give them his thoughts and opinions.

Childhood Activities and Education

I began collecting stamps when I was about seven after my father received a letter with an interesting stamp from a colleague in Trinidad. I learned a great deal about geography, governments, and history from my collection. During World War II, I did my small part for the war effort by collecting crushed tin cans so they could be recycled. I also circulated several blocks of our home with my wagon, collecting old newspapers for recycling.

My father had a workbench with many tools in the basement of our home in Arlington; he built me my own workbench. During the war, my parents shopped for bushels of vegetables and potatoes at the Southwest Market in Washington, a very poor African American area that was later developed into L'Enfant Plaza. My mother canned fruits and vegetables over gas burners in the basement. I used the burners to melt lead I collected and then made molded toy soldiers. My father worked shifts at the airport, and I often had to keep quiet at home when he returned from the midnight-to-eight or the four-to-midnight shift.

In 1943, I walked to the first grade at Nelly Custis Public Elementary School about six blocks from our home in Arlington. I didn't care much for my first teacher, Miss Holt, who criticized my sloppy penmanship, which it indeed was and still is. I came into my own in the fourth grade with an excellent teacher, Ms. Griffith. My skill in math likely began then. My interest in geography and history was fueled by my growing stamp collection.

When I was around nine, I was allowed to take the bus by myself into Washington, D.C., about five miles away. The bus stopped across from the main U.S. post office. For about seventy cents, I was able to purchase several new issues of stamps at the post office's Philatelic Department on an upper floor and have lunch.

When I was around ten, I became the "tour director" for family members visiting from out of town. We made stops at the Lincoln and Jefferson Monuments, the U.S. Capitol, the botanical garden, Pan American Union, the Tomb of the Unknown Soldier, the Custis-Lee mansion, Mount Vernon, and old parts of Alexandria, Virginia. I likely chattered. I recall visiting the U.S. Naval Academy in Annapolis with air-traffic controller Elizabeth Keep, a friend of my father.

Ethel Sykes Guildford was the oldest of my father's siblings. She took a keen interest in me because she had no children and I was the only nephew or niece in the family until after World War II. Aunt Ethel and her husband, Edgar, worked for the Veterans Administration and took trips throughout the West, sending me wonderful postcards. One had an attached small bag of salt from the Great Salt Lake in Utah. When I was in the fourth grade, she sent me a chemistry set for Christmas. I was delighted; my mother was shocked, convinced I would either poison myself or set the house on fire. My experiments fueled my interest in science, especially chemistry. Ethel always said she was the black sheep of the family. I learned much later that after attending dieticians' school in Battle Creek, Michigan, she returned to Vermont and married a local farmer. She then ran off with the pastor, Edgar Guildford, which I am sure was the talk of Brownsville in the early 1930s. Neither my parents nor my aunt Ruth would quite own up to what had happened.

After renting in Arlington, my father decided to build a home in 1948. He was 39 years old, a year older than I was in 1975 when I bought and began restoring an old home in Palisades, New York, where I still live. My parents new home was situated on a half-acre lot on a dirt road near Annandale, Virginia, then deep in the piney woods, although only about 12 miles (19 kilometers) from Washington, D.C. Now inside the D.C. Beltway, the home my father built is surrounded by thousands of houses. I was in the fifth grade at the time. I helped build the Annandale house, acquiring skills that were useful when I became a homeowner decades later.

We did not have a television set in Annandale. It wasn't until I was to be on a science program sometime in 1980 that I purchased one. In retrospect,

I wish that I had seen the lunar landings in 1969, but I was able to listen to them on BBC radio when I was recording earthquakes on remote Anegada, a British Virgin island, in the northern Caribbean.

I attended Lincolnia Elementary School for the sixth and seventh grades. While I was in the sixth grade, Wayne Lynch and I represented our school on *Quiz-Down*, a program on a local radio station featuring questions to students. Because our county did not have middle schools, we went straight from elementary school into the eighth grade at Fairfax High School. We were grouped according to ability based on recommendations by the elementary-school teachers, our grades, and our test scores. My homeroom teacher, John Waller, told us that we were the best class out of about twenty eighth-grade homerooms but that some of us would drop out and others would gain great success in life. When each of us was asked to report on a book we were reading, I described *Discovery of the Elements* by Mary Elvira Weeks (1935) to groans from the class.

In tenth-grade biology, I won a prize for an essay on natural resources. I loved chemistry my junior year and went on to make A's in freshman chemistry at MIT. I was privileged to have Clayton Taylor as a high school teacher for both world history and U.S. government. A high-level appointed official in the U.S. Department of Agriculture during the Truman administration, he came to our high school to teach after Eisenhower was elected president. Ray Williams taught me physics, trigonometry, and solid geometry my senior year. He encouraged top students and took two of us to the annual Virginia state physics exam in Williamsburg and was delighted when my classmate Andrew Gray won first place and I second. I graduated first in my senior class of about 250 students and was a member of the National Honor Society. I won the Good Citizenship Award at high school graduation, an award I still prize greatly.

I attended segregated public schools from the first to twelfth grades. Our high school in Virginia had no African Americans, Hispanics, or Asians and only a few Jewish students. Our high school sports teams were named the Rebels. The school retained other traditions left over from the Civil War, and the school system was very unequal. Mary Katherine Kern, who taught English and supervised the honor society, told me she had visited the black school for Fairfax County, a long bus trip away in the next rural county farther west, and found that the students did not have pencils and the teachers didn't even have chalk.

When I was in the tenth grade, Washington, D.C., desegregated its schools. The Fairfax County School Board then forbade our teams to play high schools in Washington, D.C., as the teams had done previously, ruling that we could play only all-white schools. Fairfax County grew rapidly during the next 50 years and, along with adjacent Arlington and Loudon Counties, voted for Barak Obama and Hillary Clinton. My mother was a census taker in 1950. She had to visit every home because there was no census by mail. She was greeted very hospitably by the members of the small African American community of Mount Pleasant on Columbia Road near Annandale.

During my junior year, I scanned college catalogs intently. I did very well on the Scholastic Aptitude Tests and applied to MIT, Rensselaer Polytechnic Institute, and Duke. Duke invited our high school student council president, Robert Dalton, and me to come to its campus to compete for a regional scholarship. Neither of us won, but we were offered small partial scholarships, which both of us declined. I applied for and received a Naval Reserve Officers Training Corps (ROTC) scholarship at Rensselaer Polytechnic. If I had accepted it, I would have been required to serve four years in the navy after graduation and likely would have missed being involved in discoveries at Lamont leading up to plate tectonics. My mother said, "Why don't you go to GW [George Washington University in D.C.] and live at home?" I was glad I left home for MIT.

MIT interviewed me at the Cosmos Club in Washington, D.C., and awarded me a full four-year scholarship. Tuition at MIT was among the highest in the country, $900 my freshman year and rising to about $1,500 my senior year. I probably would not have been able to attend without a scholarship. MIT was my first choice and, in retrospect, an excellent decision for me. Boston and Cambridge were intellectual centers and about the right-size cities for me at the time.

MIT, 1955 to 1960

Arriving at MIT as a freshman in 1955, I found views of the Charles River and the skyline of Boston beautiful. Downtown Cambridge was rather bleak, but it has then been transformed from chocolate factories and old industrial buildings to high-tech corporations and new MIT buildings.

I arrived a week before classes for rush week at MIT's twenty-five fraternities and interviewed about ten. Phi Sigma Kappa invited me to become a member, and I entered a pledge class of eight. About a third of MIT students were housed in fraternities, mostly in Boston across the river from MIT. I usually traveled by foot over the Mass Avenue Bridge twice a day from my fraternity in Boston to MIT. The cold wind blowing across the bridge was fierce in the winter. I enjoyed Boston, took many walks, and attended several performances at Symphony Hall.

I shared study and living space with two to four different fraternity brothers each semester. We didn't sleep in those rooms but in a very cold bunkroom on the top floor, where we wore heavy pajamas and warm hats, a sight to behold. I managed to survive on about five hours of sleep a night. The experience of living at the fraternity for four years was very important to me. As an only child, I never had the kinship of brothers.

Unlike most fraternities elsewhere, those at MIT involved serious study at least six days a week. We knew we had to work very hard. Nevertheless, few students flunked out. Pledges at the fraternity cleaned the house on Saturday afternoons during their first semester. We ate breakfast, lunch, and dinner at the fraternity except on Sundays and holidays, when the cook and butler were off. Cocktails and dinner on Saturday nights and sometimes on Fridays were coat-and-tie affairs, often with dates, whom we serenaded with a repertoire of fraternity songs.

All freshmen at MIT were required to take physics, chemistry, and calculus. Every Friday morning we took a written exam in one of them. But I was also fortunate that MIT had an active program in the humanities. During our first two years, we took one course per semester that was essentially devoted to great books, from Plato to Marx. My interest in the humanities blossomed. My third year I took "History of Philosophical Thought" from an inspiring teacher, Karl Deutsch, who had emigrated from Czechoslovakia before the war. He later became a Sterling Professor at Yale.

My freshman year I took "Introduction to Earth Science," an elective subject. It was taught by a new assistant professor, William Brace. I decided that year to major in the Department of Geology and Geophysics (now Earth, Atmospheric, and Planetary Sciences) with an emphasis on geophysics. Brace was my main mentor and adviser. I took his beginning and advanced courses in structural geology. His work on rock friction later influenced my graduate school work on the physics of earthquakes. When

FIGURE 2.1 Lynn Sykes, MIT, 1956.

I was a junior in 1958, Brace drove me to Washington, D.C., for the then small AGU annual meeting. I attended the only session in tectonophysics, which consisted of six or eight papers. Two of the best talks were by scientists from Lamont.

Brace took me to visit Bell Labs in New Jersey and after my freshman year in 1956 helped me to get a summer job with the U.S. Geological Survey (USGS) in Eugene Robertson's rock mechanics lab in Silver Spring, Maryland. My first professional paper in 1961, "Experimental Consolidation of Calcium Carbonate Sediment," with Robertson and Marcia Newell, resulted from that work. It was a small contribution to a much larger study by Preston Cloud of calcium carbonate deposition in the Bahamas. Robertson took me with him to Francis Birch's high-pressure lab at Harvard, where he had been a Ph.D. student. We also went to Penn State and to western Rhode Island to obtain a sample of the granite used by many workers in rock mechanics. Brace, who died in 2012, was very proud of my subsequent work in plate tectonics even though he had advised me to go to Harvard for a Ph.D.

I majored in the honors program in geology and geophysics at MIT, where I obtained BS and MS degrees in five years, finishing in 1960 (see Sykes 1960). I took many courses in math, physics, and electrical engineering as well as an advanced geophysics class on the earth's deep interior that involved much thermodynamics, geochemistry, and deep-earth structure. I attended MIT's geology summer camp at Crystal Cliffs near Antigonish, Nova Scotia. It was my first trip outside the United States. I worked with MIT graduate students who were mapping old, deformed rocks nearby for their Ph.D. dissertations. It was only with the advent of plate tectonics that the context of how those rocks formed started to emerge.

The next summer I worked for Geophysical Services Inc., soon to become a subsidiary of Texas Instruments. After ten days of lectures in Dallas, I traveled by train to Medicine Bow, Wyoming, to work with a field party on a gravity survey across a steeply dipping reverse fault. Passenger trains no longer run on that line, which is unfortunate because it traveled through great scenery in western Texas, southeastern New Mexico, and Colorado. This was my first trip west of Ohio. I finished working the rest of that summer at Shamrock in western Texas, deep in the Bible Belt. Geophysicists who worked on those field surveys lived a nomadic life, often in trailers. It wasn't for me.

MIT had an active program of lectures by well-known professors from nearby institutions, which I often attended. Their focus on international and national affairs appealed to me. I realized in retrospect that many of these professors, such as Henry Kissinger from Harvard and Walter Rostow from MIT, were strongly conservative. Some were involved in subsequent U.S. administrations and had profound effects upon the war in Vietnam and on national security policies. Kissinger had just written *Nuclear Weapons and Foreign Policy.*

MIT required all male undergraduates to take two years of ROTC. I took Air Force ROTC for two years but did not go on for an additional two. If I had done so, I would have received a commission in the air force and served for at least three years. The air force had a program in monitoring nuclear testing using geophysical methods, but it could not guarantee that I would be assigned to that program despite my training in geophysics. I likely would have become a pilot instead. Fortunately, I already had financial support for my four years at MIT and did not need additional funding.

I worked on campus as a volunteer with the Technology Community Association, served as its president my senior year, and headed MIT's Activities Council. That work likely helped in my being given the Edward John

Noble Leadership Award for graduate study—my fifth year at MIT and my first two at Columbia University. Noble had bought the ailing Life Saver brand and turned it into a financial success with good packaging and advertising.

During my fifth year at MIT, I rented an apartment on Beacon Street in Boston with two fraternity brothers. I traveled by bicycle to Harvard that year for courses in phase equilibria in mineral systems and igneous petrology and in thermal problems by Francis Birch. I found that the courses at Harvard as well as those at MIT and a combined bachelor's and master's thesis were too much for me to handle. I should have devoted more time to Birch's course and dropped the others.

Frank von Hippel, an MIT classmate, is a professor at Princeton and works on arms control, energy, and other public-policy issues. About twenty years after graduation, when I discovered he worked on energy policy, I invited him to give a lecture at Lamont. In turn, I visited Princeton about once a year to talk with Frank and his colleagues, frequently giving a lunchtime seminar on some aspect of either arms control or the safety of nuclear-power reactors.

I was a member of the board of the Federation of American Scientists (FAS) for several years when von Hippel was chair. FAS defends science, especially issues related to the nuclear-arms race. In 1986, the FAS awarded Charles Archambeau, Jack Evernden, and me its Public Service Award for "leadership in applying seismology to the banning of nuclear tests, creative in utilizing their science, effective in educating their nation, fearless and tenacious in struggles within the bureaucracy."

Woods Hole Oceanographic Institution

I received a summer research fellowship at Woods Hole Oceanographic Institution in 1959. Brackett Hersey, a marine geophysicist, invited me to work on the research vessel *Chain* for about six weeks in the Mediterranean and North Atlantic. Except for my time in Canada, this was my first trip abroad. Hersey arranged that a scientist at Hudson Labs, then a part of Columbia University, would travel with me, pay for my hotels en route, and purchase my air tickets from Paris to Messina, Sicily.

We flew on a four-engine propeller plane of the U.S. Military Air Transport from McGuire Air Force Base in New Jersey to Paris with stops in Newfoundland and Scotland, an arduous eighteen-hour journey. We spent five enjoyable days in Paris, Rome, and Messina. We had allowed additional days to make sure we reached the ship on time in Messina. My travel companion insisted on booking first-class flights for us on Air France from Paris to Rome on its newly inaugurated Caravelle jet service. The plane was luxurious, making a clear, spectacular flight over the western Alps at about 30,000 feet (9 kilometers), my first on a jet.

The research ship *Chain*, which we boarded, shot high explosives to measure seismic arrivals through the crust and uppermost mantle beneath the western Mediterranean, took measurements of deep water, obtained sediment cores, and made bottom soundings with a precision graphic recorder (PGR). I took shifts of about six hours overseeing the PGR. The U.S. Navy paid for much of the cruise because it involved many measurements of underwater sound. The cruise was leisurely, especially by Lamont standards. It stopped at Monaco, where Jacques Cousteau gave us a reception; Barcelona; and the naval base at Rota, Spain.

Hersey was interested in strong seismic reflections that bounced back to the ocean's surface from layers within the shallow sediments. These reflections were detected on the PGR and seemed to originate at the boundaries between sand and clay layers in cores we took in the Tyrrhenian Sea. Each core consisted of sediments extracted from a pipe about 30 feet (9 meters) long that was driven into the ocean floor by a 500-pound (225-kilogram) weight. Hersey said a prominent British acoustician gave him a set of acoustic probes several inches long. He suggested I use them to make measurements of sound velocity along some of the cores as part of my BS/MS thesis. I agreed to work on those cores after the cruise at Woods Hole and then in the soil mechanics lab at MIT.

After working on the sediments from one core at MIT for nearly a semester, I found that the layers in the core were not horizontal, as they were originally, but were a chaotic mixture known as a "suck in." It was clear that this core had not been obtained with care by other scientists on the ship and so was useless for scientific study. Maurice Ewing and others at Lamont would not have allowed that to happen because Lamont ships took about two cores per day and many scientists worked on them. Although I was then able to use other undisturbed cores, the probes Hersey gave me were poorly constructed and a poor choice for my experiment.

Hersey complimented me on my work, and we submitted a paper to the journal *Geophysics.* Soon after I arrived at Lamont in the fall of 1960, however, the paper was rejected as being too long and more like an undergraduate thesis, which in retrospect it was (see Sykes and Hersey 1960). Hersey blamed me, but he should have given me more help in editing and shortening it before we submitted it to a journal. I vowed never to let that happen with any of my own students.

Helsinki and the Earth's Free Oscillations

During the summer of 1960, I sailed on the Woods Hole–operated ship *Chain* from Bermuda to Helsinki, Finland. We had a spectacular voyage around weather-beaten Cape Wrath at the northwestern tip of Scotland and then across the North and Baltic Seas. The ship was on exhibit at Helsinki for a week at the meeting of the International Union of Geodesy and Geophysics. Ewing would never have permitted a Lamont ship to spend such a large amount of time in port just for display. When asked "Where do you keep your ships?" he is said to have replied, "I keep my ships at sea." Lamont ships were at sea for eleven months of the year, largely in the stormy southern oceans.

In Helsinki, I attended several sessions on seismology and heard Keith Bullen and Sir Harold Jeffreys speak for the first time. As applied mathematicians, they had derived travel-time tables of seismic waves that were used globally to locate seismic events and to understand deep-earth structure.

Especially important to me were the initial results of recordings of the earth's free oscillations that were generated by the giant Chilean earthquake of May 1960. This earthquake was the largest recorded shock to date since seismographs were invented in the 1890s and still is as of late 2018. The idea that the whole earth could ring like a bell at various tones (frequencies) had been proposed but never recorded unequivocally until May 1960.

Although the session in which these results were reported was originally scheduled to be on long-period seismic waves, several authors—including those from Lamont, Cal Tech, UCLA, and the Weizmann Institute of Israel—changed their presentations to describe data and calculations for the giant Chilean earthquake that had occurred just three months earlier. Well before email and the Internet, presenters had to rely on colleagues at their home institutions to send updated information by telegram. One of the earth's free oscillations, called the "football mode," rang like a bell but with a very long period, taking fifty-three minutes to complete just one cycle of vibration. It was one of the most exciting scientific sessions of my life.

On my Pan American flight home from Helsinki to New York, I introduced myself to Joe Worzel, the associate director of Columbia's Lamont Geological Observatory. He asked, "Why didn't you stay on ship for the entire cruise?" Doing so was typical for new students in marine sciences at Lamont but not for students at most other institutions. It would have meant six to twelve months at sea. By then, however, I had visited Jack Oliver at Lamont and decided that seismology was likely to be my main field of interest. In addition, seasickness had been a major problem for me on the two *Chain* cruises.

New York City, Columbia University, and Lamont

I was ready to leave MIT at the end of my fifth year. I asked Robert Shrock, the head of the Department of Geology, for his recommendations on Ph.D. programs in geophysics. He said, "Stay at MIT or go to Cal Tech." Nevertheless, I already had read articles in professional journals and had noticed many exciting publications by Lamont scientists. When I asked him, "What about Lamont?" he replied, "It's good too." By the middle of my fifth year at MIT, I applied and was accepted at Berkeley and Cal Tech. When I visited Lamont on a Saturday during the spring of 1960, Jack Oliver, a professor and seismologist, spent a lot of time with me and said I would be admitted if I applied.

I soon decided to attend the Department of Geology at Columbia and to do my research at Lamont with Oliver as my adviser. That department is now called the Department of Earth and Environmental Sciences, and Lamont is now the Lamont-Doherty Earth Observatory. The department is the degree-granting part of the earth sciences. Most of the department's faculty members as well as many scientists at the Ph.D. level do their research at Lamont.

Ph.D. programs were open to greater numbers of Americans by 1960. I was surprised by my mother's response when I told her I was going to work on a Ph.D. She replied, "Our kind of people don't go to graduate school." Perhaps she thought I would either make little money as an academic or face barriers and disappointments because of my middle-class upbringing.

I moved to New York in 1960 when I began graduate school at Columbia, renting an apartment on West 153rd Street in Manhattan. It was not a great neighborhood, but I could afford the $69 a month. In 1965, after receiving my Ph.D., I moved to West 77th Street near Riverside Drive, a much nicer neighborhood. I had a one-bedroom apartment in a renovated brownstone designed by architect Clarence True in the 1890s. My salary at the time had jumped to $8,500 per year, but even then I wondered if I could afford the rent of $220 a month. Fortunately, my salary at Columbia continued to rise.

In 1975, when I was the president of the Far-West 77th Street Block Association, we had an all-day party to raise money to plant curbside trees. It's good to see them fully grown when I drive by today. We republished a small book by Clarence True, who had designed about one hundred residences in the neighborhood. I became well acquainted with art galleries, museums, opera, ballet, and Central Park in New York City and took many long walks.

After a half-year sabbatical in China and Japan in 1974, I started looking for a house near my office at Lamont. I found an old historic home close by in Palisades, New York. Originally built in 1865, it was in bad shape and required considerable work. It was more than a "fixer-upper." For more than a year, I spent much of my free time restoring it. The wide plank floors in four rooms and the front hall were beautiful once I removed many coats of ugly old paint. The inside and outside had not been painted in at least 25 years. The house needed a new roof down to the rafters. Once I took possession, my father came to work on it with me for a week. I decided not to repaint inside moldings after they had been stripped but to stain them instead, recalling that Japanese aesthetic from my stay in Japan during late 1974.

After 40 years, I retired from Columbia and Lamont in 2005. I now work mainly at home in the Palisades on research papers and books and walk to Lamont about twice a week for seminars, earthquake readings, lunch, and talks with other scientists.

3

EARTHQUAKES ALONG FRACTURE ZONES AND MID-OCEANIC RIDGES, 1963-1965

Lamont Geological Observatory of Columbia University

I arrived at the Lamont Geological Observatory of Columbia University in September 1960 as a graduate student working toward a Ph.D. in seismology. During my first year at Lamont, I spent about half of my time taking classes on the main Columbia campus in Manhattan and the other half working on a research project in seismology at Lamont under my adviser, Jack Oliver. I had an office in a former bedroom with four other people on the second floor of Lamont Hall, with a beautiful view up the Hudson River.

Lamont Hall was formerly the home of Thomas Lamont, the senior partner in the J. P. Morgan bank. After he died in 1949, his wife offered the Lamont estate of about 3,000 acres to Columbia University. Columbia was loath to accept estates because considerable endowment was needed to maintain them. The Department of Geology at Columbia hired Maurice "Doc" Ewing as a new, lone geophysicist after World War II. He assembled a group under him that first worked on studies of the earth beneath the oceans in a few rooms in Schermerhorn Hall on the main campus of Columbia. Ewing had obtained federal funding for this work as a result of his important contributions to ocean acoustics and submarine detection at Woods Hole during the war. He wanted much more space than was available on the main campus and needed a quiet location to operate seismographs, so he persuaded the president of Columbia, Dwight Eisenhower, to accept the Lamont gift, which Eisenhower did in 1949. Ewing then became the first director of what then became the Lamont Geological Observatory.

Ewing was very much the person in charge of most early research at Lamont. With the exception of geochemistry, he had his fingers in most projects. His name was on many published papers. For some of them, he was greatly involved, but for about two-thirds his input was minimal. Many of Ewing's early students became senior faculty members and went on to direct major projects as Lamont grew in the 1950s. They included Jack Oliver and George Sutton in seismology, Bruce Heezen in marine geology, and Charles Drake in marine geology and geophysics. Ewing had trouble giving up his scientific and administrative involvements to these and other distinguished scientists.

Probably one of the most important learning experiences for seismology students at Lamont was a weekly seismology seminar. On Monday evenings, either people on the staff or graduate students gave an hour presentation, including questions and slides, on their latest work. An important legacy from both Ewing and Oliver was the importance of picking a topic that was ripe for really important research. They believed that you didn't have to be a genius to make an important discovery. That turned out to be true many times at Lamont. Several very smart people elsewhere worked on second- and third-rate problems.

Oliver also emphasized good writing. Even in the 1960s so much was written in geophysics and geology that if you wrote a poor introduction, people would not read the rest of the paper. Oliver emphasized an introduction that states why the paper is important, summarizes what is to follow, and then outlines the main conclusions. I continue to follow his advice.

I later worked with each graduate student of mine on the scientific organization of manuscripts, good writing, and clear figures. Usually I went over several versions of a manuscript with a student before considering it ready for submission to two Lamont reviewers and then to a scientific journal. I did not want my students to have the negative experience with a manuscript that I had with Brackett Hersey at Woods Hole.

The International Geophysical Year from 1957 to 1958 provided the opportunity for scientists to operate and analyze earth science data from many sites around the world. Lamont was instrumental in installing about a dozen sets of long-period seismographs around the world, which were developed and tested at Lamont. This was an important precursor to the establishment of the WWSSN stations mentioned in chapter 1. By 1960, original records were arriving at Lamont from those stations—some once a year, others every two weeks. Oliver was very interested in looking at records just to see what new and exciting things they might contain. I got into the habit of doing that, too, learning about the types of seismic waves recorded at various distances and looking for new phenomena. I probably have looked at as many seismograms as any other seismologist.

One of my officemates in my first year at Lamont, Lee Alsop, had recently received his Ph.D. in physics at Columbia under Charles Townes. Alsop was working on the giant Chilean earthquake of May 1960, which was well recorded by the long-period seismic stations that Lamont had installed a few years earlier. Only a month before the giant earthquake, Lamont had also installed a long strainmeter in a deep mine at Ogdensberg, New Jersey. It was made of fused quartz, fixed at one end to rock, and connected to a sensor (capacitor) that measured ground displacement at the other end. Strainmeters are more sensitive than seismographs to very long-period waves like those generated by the Chilean shock. Since the cement piers at the ends of the strainmeter at Ogdensberg were still curing and hence deforming, that strainmeter recorded the great earthquake and soon went off scale. We were lucky the event was recorded. Alsop supervised the digitizing of the analog (wiggly line) Chilean records so he could analyze their spectral content and identify the earth's free oscillations. This was decades before most analog seismic recordings were replaced by digital recording.

Work on Seismic Surface Waves for My Ph.D.

Throughout the 1950s and 1960s, many seismologists at Lamont and Cal Tech conducted research on seismic surface waves. Unlike the faster P and S waves that travel deep into the interior of the earth, *surface waves* typically have their largest motions (amplitudes) near the earth's surface (figure 3.1). The two main types of surface waves are called *Love* and *Rayleigh waves* after two British scientists who predicted their occurrence theoretically. Love waves travel about 10 percent faster than Rayleigh waves. Earthquakes generate both types of surface waves, but large explosions typically produce much larger Rayleigh waves than Love waves. For my work, I used both types of seismic surface waves.

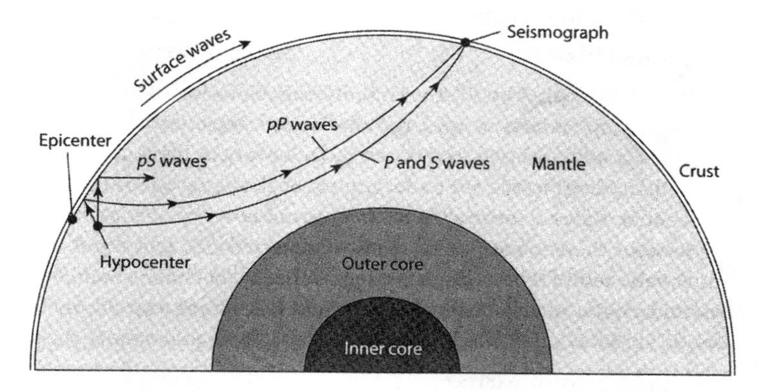

FIGURE 3.1 Interior of the earth. Solid lines indicate paths of seismic waves between a source, such as an earthquake, and a recording station. The *epicenter* is the point on the earth's surface directly above the source or *hypocenter* of an event.

Source: Sykes and Evernden 1982.

My first paper at Lamont, "Mantle Shear Wave Velocities Determined from Oceanic Love and Rayleigh Wave Dispersion," was published in 1962 with Lamont graduate student Mark Landisman and visiting Japanese seismologist Yasuo Sato. Our paper sought to fit seismic recordings of those types of two seismic surface waves by using a layered spherical earth model for its outer 300 miles (500 kilometers). Although computations of that type are commonplace today, they were new in 1962. Most previous calculations for a spherical earth had been performed with a series of flat homogeneous layers with what was called the "earth-flattening approximation." Our results for Love waves on a sphere differed from those done with earth-flattening computations. We were able to resolve why earlier computations with flattening approximations gave different (and incorrect) results for Love waves.

Earthquakes Along Enigmatic Fracture Zones

Here, I would like to take a step back in time from my work on transform faults, described in chapter 1, to earlier work on earthquakes along what were then called *fracture zones* and those along the crests of the Mid-Oceanic

Ridge system. Prior to 1966, fracture zones were described as enigmatic, very long, linear zones of rough topography on the seafloor, some of which intersected Mid-Oceanic Ridges. My work on transform faults came about indirectly through my examination of surface waves generated by earthquakes in the southern Pacific Ocean. My road to plate tectonics, continental drift, and seafloor spreading was rather serendipitous.

In 1961, I decided to examine certain types of seismic surface waves that were of high frequencies (short periods of four to twelve seconds). They propagate across oceanic areas and are sensitive mainly to the properties of sediments on the ocean floor and the thin oceanic crust below. That work needed long paths from earthquake sources in the oceans to instruments that recorded them on either islands, such as Hawaii, or near coastlines. The data I accessed from oceanic paths in the southern oceans came mainly from the seismograph stations put out around the world by Lamont during the International Geophysical Year in 1957 and 1958. To make accurate determinations of the velocities of those surface waves, I also needed to have accurate locations and times of occurrence of the earthquakes that generated them.

I soon realized that the published locations in standard catalogs for many earthquakes in the southern oceans had uncertainties of 100 miles (160 kilometers) or larger. In addition, most of the earthquakes of interest to me had been located using the times of arrival of the first seismic P waves at only a few stations. Figure 3.1 shows the paths P waves travel in the deep interior of the earth. Prior to the determination of earthquake locations using computers in the early 1960s, including more than a few P wave readings in a study was very labor intensive. The locations of older shocks were often determined crudely using tape measurements on a globe.

To determine more accurate locations, I adapted a computer program that Bruce Bolt wrote in 1960 when he was a postdoctoral scientist at Lamont. His program reduced the uncertainties in calculating earthquake locations. He input a large number of P wave arrival times from tens to hundreds of stations into a computer, thus avoiding much intensive labor in the calculation of an earthquake's location and time of occurrence. An applied mathematician, Bolt was interested mainly in analyzing a few earthquakes to demonstrate his method, not to examine the geological setting of the earthquakes themselves.

My initial interest was merely to improve my calculations for surface waves. The geologic setting of earthquakes in oceanic areas, however, soon became much more interesting to me. I proceeded to analyze many earthquakes along the Mid-Oceanic Ridge system in the southern Pacific, where many of the shocks of interest to me occurred. Mark Landisman helped me rewrite Bolt's program for the IBM 7090 computer at the NASA Goddard Space Flight Center near Columbia University. I received time on what was then one of the fastest computers in the world—before the existence of personal computers and the Internet. Today my personal computer is more powerful than the 7090 was in 1962.

Landisman and I added a library of the latitudes and longitudes of hundreds of seismic stations around the world along with three-letter abbreviations for each of them. For each earthquake, I entered P wave arrival times and station codes onto separate punch (IBM) cards (lower half of figure 3.2). A lead card contained the initial or trial earthquake location (upper half of figure 3.2). Up to several hundred cards per earthquake were used in a computer run. Fortunately, the NASA center had very fast mechanical card readers as well as rapid printers.

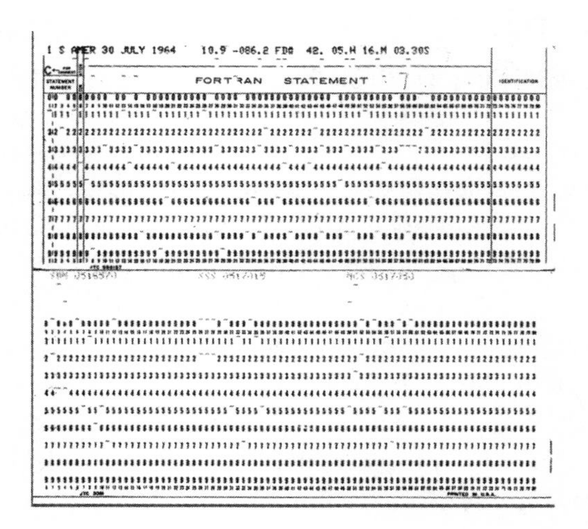

FIGURE 3.2 Two punch cards, part of a relocation of an earthquake by computer. The top card gives the earthquake's trial (initial) latitude, longitude, depth, and origin time; the lower card gives three-letter abbreviations and arrival times of P waves at three stations—SDM, SSS, and NCS.

Using the computer program to look up the location of each station via its three-letter code saved much time and effort. It calculated the distance and travel time of P waves to each station for an event using the trial location and its time of occurrence. The program then computed a best-fitting revised location, origin time, depth, and their uncertainties for each earthquake in a fraction of a second using data from all reporting stations. I was able rather quickly to compute revised locations and origin times for many of the events for my paper "Seismicity of the South Pacific Ocean" (1963), as shown in figures 3.3 and 3.4. I then used those refined locations and times of occurrence to obtain more accurate velocities of the short-period seismic surface waves that I initially set out to study.

Although the U.S. Coast and Geodetic Survey had also started locating earthquakes by computer in the early 1960s, in 1962 I was looking for a larger sample of earthquakes going back a decade earlier. Earthquake locations in the Southern Hemisphere were improved considerably when P wave arrivals became available in the 1950s from the several seismic stations in Antarctica.

A Major Fracture Zone Intersecting the Mid-Oceanic Ridge System Identified by New Locations of Earthquakes

On plotting revised earthquake locations in the southeastern Pacific (figure 3.3), I found that they were confined to narrow bands along parts of the world-encircling Mid-Oceanic Ridge system. In the eastern Pacific, the northern part the ridge system is called the East Pacific Rise, and the part south of latitude 55°S is known as the Pacific-Antarctic Ridge.

Those ridges on the seafloor stand out in figure 3.3 as white areas where water depths are shallower than about 2,000 fathoms (12,000 feet or 3,660 meters). Another band of earthquakes follows what is called the West Chile Ridge, which extends from Easter Island to southern Chile. What particularly caught my eye was the concentration of earthquakes near latitude 55°S between longitudes 122°W and 135°W (figures 3.3 and 3.4). The northeasterly trend of earthquakes along the Pacific-Antarctica Ridge to the south suddenly changes direction to easterly at 55°S, 135°W. It then resumes a

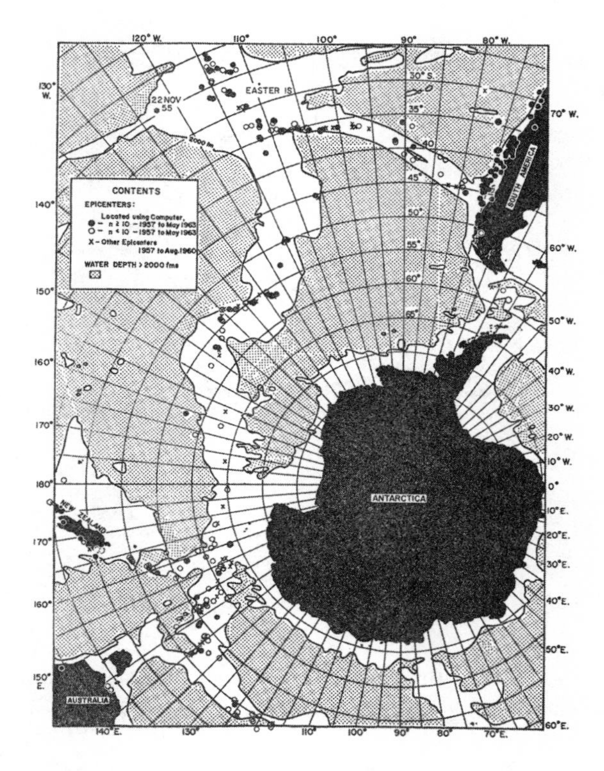

FIGURE 3.3 Revised earthquake locations in the South Pacific from 1957 to 1963. Contours of 2,000-fathom water depths are shown.

Source: Sykes 1963.

northeasterly course at 56°S, 122°W. That pattern led me to propose that a major mountainous feature on the seafloor, a *fracture zone*, trended nearly east–west and intersected the Mid-Oceanic Ridge system at latitude 55°S. I found that the seismically active part of the fracture zone between those two ridge crests was nearly 620 miles (1,000 kilometers) long. The proposed fracture zone was seismically active only between the two crests of the ridge system but not farther east or west.

In the 1950s, William Menard of the Scripps Institution of Oceanography had discovered several enigmatic features on the seafloor that intersect the Mid-Ocean Ridge system at high angles in the eastern Pacific. He described these fracture zones well to the north of my study area as long,

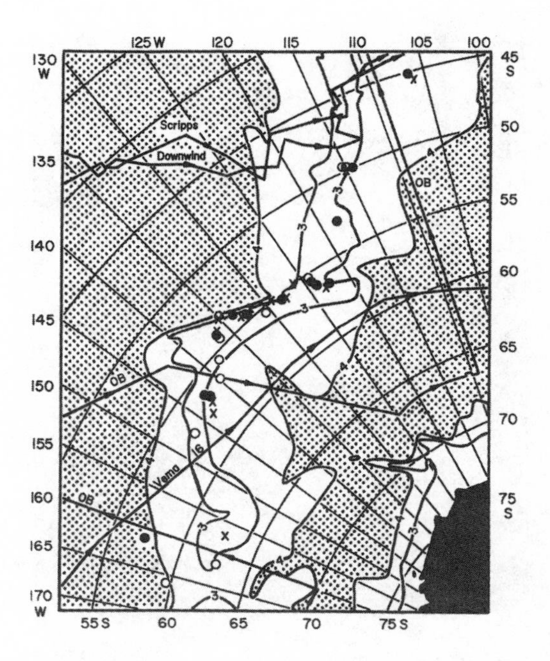

FIGURE 3.4 Enlarged view of earthquakes beneath the southeastern Pacific Ocean. Ship tracks and water depths of 1.9 and 2.5 miles (3 and 4 kilometers) are labeled.

Source: Sykes 1963.

linear zones of rough topography. It is understandable that he did not see another zone located near 55°S because my proposed zone straddled the "pole of inaccessibility" near the farthest point from inhabited land on Earth. Ships infrequently surveyed this region. That area also is typified by some of the roughest seas and the most severe weather on Earth.

Menard reviewed my manuscript prior to publication and sent me an unpublished map of seafloor depths for my study area, which I included in my paper (Sykes 1963) with his permission. In his review, he agreed that my study area included a major fracture zone.

My discovery of an active fracture zone and the zigzag pattern of earthquakes using improved earthquake locations was fortuitous because my original intention was to study only a certain type of seismic surface wave. I was on my way to plate tectonics with my discovery of that fracture zone and its intersection with the ridge system. How could a great fault or fracture zone suddenly end? The zigzag pattern and the lack of earthquakes beyond the ridge segments was one of J. Tuzo Wilson's key pieces of evidence in 1965 for his proposal of a new class of faults, which he called *transform faults.*

When I showed my results to Maurice Ewing, the director of Lamont, he immediately called Bert Crary at the National Science Foundation and arranged for the research vessel *Eltanin* to map the region near 55°S. The hull of the *Eltanin* had been reinforced so that it could work in the rough seas and ice surrounding Antarctica. Ewing named my major feature the "Eltanin fracture zone." (Eltanin is the brightest star in the constellation Draco.)

At Ewing's suggestion, Lamont graduate student Walter Pitman sailed on the *Eltanin* to map the region in more detail. Although the original intention was to study the Eltanin fracture zone, Pitman was drawn to variations in the earth's magnetic field, called *magnetic anomalies*, somewhat farther to the northeast. In 1966, Pitman and James Heirtzler, who headed the magnetics group at Lamont, published their "magic profile" of magnetic anomalies across the Ridge. Its interpretation in terms of seafloor spreading and continental drift is described in chapter 1. Pitman might not have gone to sea on the *Eltanin* or obtained his "magic profile" if I had not proposed in 1963 that a great fracture zone intersected the ridge system along the southern East Pacific Rise.

The magnetic anomalies described by Pitman and Heirtzler were obtained along a ship's track that crossed the East Pacific Rise. The simplicity of that magnetic-anomaly pattern likely resulted from the fast rate of seafloor spreading (opening) along that segment of the Mid-Ocean Ridge system. It is far from any continental or other disturbing geological feature. Later mapping revealed that the Eltanin fracture zone consisted of three closely spaced faults that then were named for former Lamont scientists Bruce Heezen, Marie Tharp, and Charles Hollister.

With my interest in seismic surface waves in oceanic areas and with my finding of a great fracture zone in the far southeastern Pacific, I regularly read papers on marine geology and geophysics by a dozen Lamont scientists. I talked with many of them on a regular basis and kept in close touch with the work of my colleagues in seismology—Lee Alsop, James Brune, James Dorman, Bryan Isacks, Mark Landisman, Peter Molnar, Paul Pomeroy, Jack Oliver, Christopher Scholz, and George Sutton. My interactions with my colleagues at Lamont increased in 1966 when I confirmed Wilson's concept of transform faulting.

Through these contacts, I was able to gain greater knowledge as well as encouragement as I worked on other fracture zones and the distribution of earthquakes beneath the oceans and then of those at island arcs such as Tonga, Kuril-Kamchatka, Japan, and the Aleutians. Island arcs are long, curved chains of oceanic islands and their parallel offshore deep-sea trenches. They are the sites of intense volcanism and earthquake activity that extend from the earth's surface to great depths. They were called *subduction zones* after 1968.

In looking back, I realize that if I had gone to graduate school at either Cal Tech or Berkeley and had not been at Lamont, I would not have been involved in the discoveries of transform faulting, seafloor spreading, continental drift, subduction, and, finally, plate tectonics. Neither of those two institutions nor MIT had much involvement with marine geology or the subsequent development of plate tectonics. Lamont, Princeton, Scripps Institution of Oceanography in the United States, Toronto University in Canada, and Cambridge University in the United Kingdom were the main contributors to plate tectonics.

My Identification of Other Great Fracture Zones in the Oceans

Finding a previously unknown fracture zone that was seismically active and nearly 620 miles (1,000 kilometers) long was a very exciting discovery to me. I quickly realized that it was more important than my results on seismic short-period surface waves. Working to finish my study on surface waves for my Ph.D. dissertation, I decided to try concurrently to identify other major fracture zones in the oceans from their seismic activity. I also decided to determine more precise locations of earthquakes at several island arcs to better define the configuration of their inclined seismic zones, some extending to depths as great as 430 miles (690 kilometers), as described in chapters 4 and 5.

In 1964, Landisman and I relocated a large number of earthquakes in East Africa, the Gulf of Aden, and the Arabian and Red Seas (see Sykes and Landisman 1964). A year earlier Drummond Matthews of Cambridge

University described a major fracture zone near the mouth of the Gulf of Aden in the Arabian Sea that appeared to offset the Carlsberg Ridge, a branch of the Mid-Oceanic Ridge system in the northwestern Indian Ocean.

As in the southeastern Pacific, we found that seismic activity was confined to a narrow zone that trended east–northeast along the center of the Gulf of Aden, zigged to the south–southwest along that fracture zone, and then suddenly switched to a southeasterly trend along the Carlsberg Ridge. Once again, most of the activity along the fracture zone was confined to the space between two ridge crests.

High seismic activity occurred where other horizontal faults (transform faults) trending north/northeast–south/southwest intersect the median (central) ridge in the Gulf of Aden. By 1961 and 1964, Heezen and Tharp of Lamont also had identified a number of fracture zones on their new maps of the seafloors of the South Atlantic and Indian Oceans.

In 1965, I relocated many well-recorded earthquakes that occurred between 1955 and early 1964 in the Arctic Ocean and the Norwegian Sea. I found a narrow zone of earthquakes that trended northeasterly to the north and south of Iceland along a segment of the Mid-Atlantic Ridge. The seismic zone north of Iceland in the Norwegian Sea suddenly changed direction to an easterly trend near the volcanic island Jan Mayen at latitude 72°N. I called this 300-mile (500-kilometer), easterly trending feature the *Jan Mayen fracture zone*. Again, the relocated earthquakes along it occurred only between two segments of the Mid-Atlantic Ridge, not farther east or west. Near its eastern end, the pattern of earthquakes gradually changed direction to northeasterly and then north/northwesterly farther north to the west of Spitsbergen (Svalbard). The Jan Mayen zone was subsequently mapped as a series of closely spaced transform faults.

In my paper "The Seismicity of the Arctic" (1965), I also mapped a nearly straight, narrow belt of earthquakes that extended across the eastern (Eurasian) basin of the Arctic Ocean from off the northeast coast of Greenland to near the mouth of the Lena River in northeastern Siberia. I reported that the earthquake belt is parallel to and very nearly equidistant from the Lomonosov Ridge in the center of the Arctic basin and the northern edge of the continental shelf of Eurasia. (Earth scientists often use the name "Eurasia" for Asia and Europe combined.) The earthquake belt is an extension across the Arctic of the crest of the active Mid-Atlantic Ridge, but it did not follow the Lomonosov Ridge as some Russian seismologists and geologists had maintained.

We now know instead that the Lomonosov Ridge, which is not seismically active, was split off as a continental fragment or sliver from the northern edge of Eurasia. The seismic zone instead follows what is now called the Gakkel Ridge, named after Y. Y. Gakkel, a Russian oceanographer who discovered it as an elevated feature in the eastern basin of the Arctic Ocean. Gakkel did not realize that this ridge's elevated crest was seismically active and a branch of the Mid-Ocean Ridge system. The western Arctic basin, to the west of the Lomonosov Ridge, is not seismically active and must have formed much earlier. My work also helped to define where active seafloor extension occurred and is still occurring beneath the Arctic Ocean.

In 1964, I published the material in my Ph.D. thesis on short-period seismic waves in two articles in the *Bulletin of the Seismological Society of America* (see Sykes and Oliver 1964). Few people outside of seismology have read it. I knew then that I wanted to put my main efforts into finding additional seismically active fracture zones and examining the detailed distribution of earthquakes beneath island arcs.

In 1964, I also traveled to Bermuda to work on seismic surface waves and the T phase, which propagates very efficiently through the oceans. I used an extensive collection of data at the seismograph station in Saint George. To save money, I stayed in a room at an old building that housed the Bermuda Biological Station. I could look out and see a large burned area between the station and H.M. Prison, where a U.S. bomber tried to take off and didn't make it during the Cuban Missile Crisis in 1962. I learned that the director of the Biological Station had been arrested for poisoning his wife by serving her homemade methyl alcohol.

Partway through my work on surface waves, Ewing asked me to work on images from new weather satellites to study microseisms (earth noise) generated by hurricanes and other storms. Ewing and William Donn published a number of papers on microseisms in the 1950s in attempts to track hurricanes. They and others found, however, that strong microseisms were generated mainly when hurricanes reached continental shelves where water depths were much shallower than those in the deep oceans. Hence, microseism data arrived too late to give much warning to land areas.

I worked for a few months on microseisms, but with little success. Weather satellites, not microseisms, had already become much better tools to track hurricanes before they reached continental shelves. I knew instead that I wanted to put my main efforts into earthquake locations along ridges and beneath island arcs.

George Sutton, an associate professor of seismology, said that Ewing resented my mentor Jack Oliver for "stealing seismology from him." By 1960, however, Lamont was too large for one person to supervise and participate in its many areas of research, as Ewing still tried to do. Fortunately, I was able to return to work with Oliver on seismic surface waves. Microseisms, which did not become an important area of study again for many decades, would have been a dead end for me.

4

EARTHQUAKES AT SUBDUCTION ZONES, 1965–1967

In addition to work on relocating earthquakes along fracture zones and Mid-Oceanic Ridges, in 1965 I began refining the locations and depths of seismic events along island arcs, as they were known then but were now referred to as subduction zones. They are places where two plates converge upon one another—one riding (thrusting, figure 1.3) over the other and the other plunging deep into the earth's mantle—as in the island arcs of Japan and the Aleutians. They are the sites of abundant shallow and deep earthquakes, deep-sea trenches, and volcanoes.

Seminar on Deep Earthquakes

Jack Oliver, my Ph.D. adviser, ran several seismology seminars at Lamont for graduate students. Each semester we chose an interesting topic such as local earthquakes. In 1963, I participated in one of Oliver's seminars on deep earthquakes with fellow graduate student Bryan Isacks and a few others.

Very little work had been done on deep earthquakes during World War II and for more than 15 years thereafter. As a consequence, it was a ripe topic for good new research. Earthquakes at several island arcs extend down to depths as great as 430 miles (690 kilometers), about 10 percent of the earth's radius. In the seminar, we tried to read all known papers on deep earthquakes, something that would be virtually impossible to do today. Oliver sensed that the study of seismic surface waves was slowing and that our group needed to pursue other topics. When I arrived at Lamont, virtually everyone in seismology was working on surface waves. This particular seminar led Isacks and Oliver to write a proposal to the National Science Foundation for funds to install seismic instruments in the Tonga–Fiji region of the southwestern Pacific, the site of more frequent very deep shocks than any other place in the world (darker shaded areas in figure 1.1).

Earthquakes and the Deep Structure of Island Arcs

The seminar on deep earthquakes led me to undertake a relocation of about 1,500 earthquakes of all depths in the Tonga–Fiji area of the southwestern Pacific as well as for island arcs of the northwestern Pacific. I completed most of those relocations before I went to Fiji to work with Isacks in July 1965. Upon my return, I submitted the paper "Seismicity and Deep Structure of Island Arcs" in December 1965, using those relocations. It was published in June 1966, just in time for me to start work on mechanisms of earthquakes and transform faulting along the Mid-Oceanic Ridges.

Nearly all of the world's deepest earthquakes occur at island arcs—that is, subduction zones. The greatest numbers of those deepest shocks are concentrated between the islands of Fiji and Tonga in the southwestern Pacific. In contrast, the global Mid-Oceanic Ridge system and its extension into the East African Rifts are well defined by very shallow earthquakes (shown in figure 1.1 and in red in plate 1).

As described in the paper I published in 1966, I found that relocated earthquakes beneath island arcs of the southwestern and northwestern Pacific extended from the surface near deep-sea trenches, V-shaped depressions that are the deepest parts of the oceans and are formed by underthrusting, to the depths of the deepest earthquakes. The inclined earthquake zones (figure 4.1) generally are quite thin. Many American earth scientists called them "Benioff zones" after a well-known paper in 1954 by the Cal Tech seismologist Hugo Benioff. My colleagues and I at Lamont, however, referred to them as "Wadati-Benioff zones," knowing from our

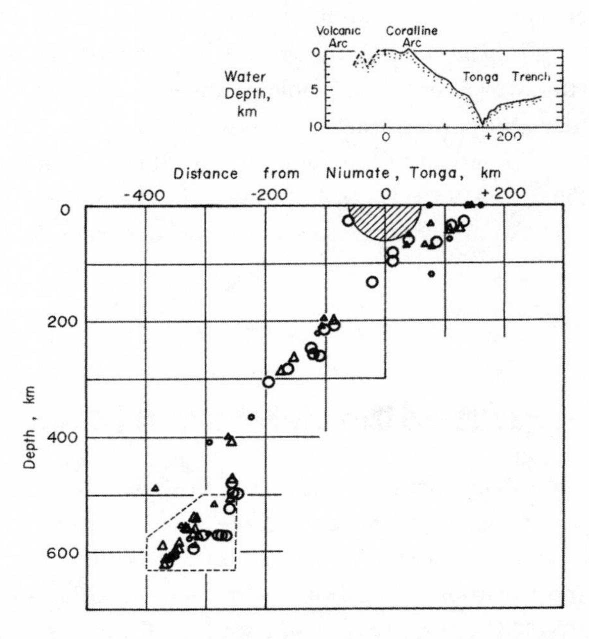

FIGURE 4.1 Vertical cross-section oriented perpendicular to Tongan subduction zone (island arc) showing the inclined seismic zone. Circles and triangles are relocated hypocenters of earthquakes. Note the small thickness (less than about 12 miles [20 kilometers]) of the seismic zone for a wide range of depths and the more diffuse zone for depths greater than about 300 miles (500 kilometers).

Source: Isacks, Oliver, and Sykes 1968.

seminar that Japanese seismologist Kiyoo Wadati had studied them earlier in the 1920s and 1930s.

One of my important findings in 1966 was that the zones of earthquakes at all depths, the deep-sea trench, and the active volcanoes curve abruptly to the west just south of Samoa at the northern end of the Tongan island arc (figure 4.2). I proposed that the tectonic processes responsible for the sudden curvature of these features must extend from the surface to the depths of the deepest earthquakes. What I did not comprehend then was that these features and processes were linked by the overall process that was soon to be called "subduction"—that is, one plate plunging beneath another into the upper mantle of the earth.

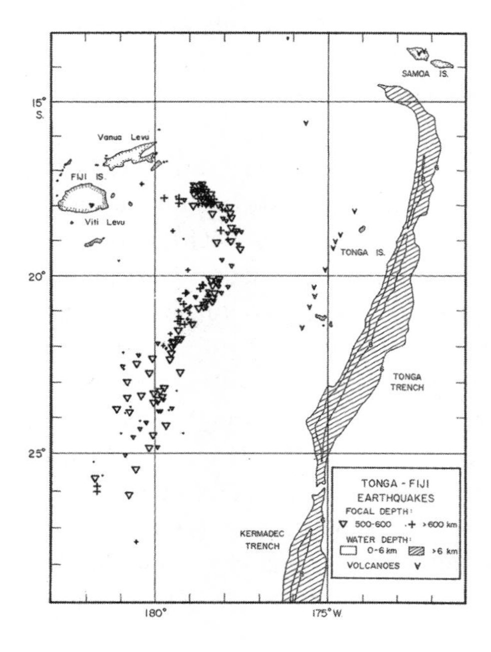

FIGURE 4.2 Very deep earthquakes beneath the Tonga–Fiji region of the southwestern Pacific with depths greater than 300 miles (500 kilometers). Note the strong hook to the northwest in deep earthquakes, volcanoes, and the Tongan trench. Shallower earthquakes (not shown) exhibit the same abrupt hook.

Source: Sykes 1966.

Continental Drift

I presented my results on revised earthquake locations for Tonga–Fiji at a symposium at Newcastle, England, in late March 1966. I stated orally but not in my abstract that such tectonic structures extending to depths as great as 430 miles (690 kilometers) were difficult to reconcile with the concept of continental drift—the movement of continents with respect to one another—which had first been proposed early in the twentieth century by German meteorologist Alfred Wegener.

Many of the papers at the symposium dealt with continental drift from the perspectives of paleomagnetism and paleoclimate. Harold Urey, a Nobel Prize winner in chemistry, said to me in private after my talk, "Young man, you should take continental drift more seriously." And, indeed, a few months later I changed my views about global deformation of the earth and started to work nonstop on continental drift and other concepts of a mobilistic earth.

Many scientists at Lamont, including Director Ewing and Associate Director Worzel, were avid believers in continental fixity throughout geological time. Ewing thought that the oldest rocks on Earth would be found beneath the deep oceans. This, of course, turned out not to be the case. Lamont ships traveled the world studying the ocean floor. Ewing expected that his exploration of the earth beneath oceans would lead him to a revolutionary understanding of the earth that differed from existing theories. Deep-sea cruises operated by Ewing and others at Lamont did provide crucial data that, in fact, confirmed seafloor spreading, continental drift, and plate tectonics—not what Ewing had expected.

Ewing was not alone. Many others at Lamont as well as most North American geologists did not accept the theory of continental drift and were too busy working on either regional or smaller-scale features even to consider it. In fact, when I was an undergraduate at MIT, a professor told me that respectable young earth scientists should not work on vague and false concepts such as continental drift: continents could not plow through the strong oceanic crust and upper mantle of the earth. When I was an undergraduate at MIT, Walter Munk talked about his book *The Rotation of the Earth*, published in 1960 with Gordon MacDonald. Munk claimed that evidence for drift was poor and that the subject should be avoided. Many rejected continental drift for the wrong reasons. It occurs, but its mechanism is not that of continents plowing through the solid crust and mantle beneath the oceans.

Neil Opdyke, who had worked as a graduate student on continental drift under Keith Runcorn at Newcastle, England, was an exception when he arrived at Lamont as a research scientist. As an undergraduate majoring in geology at Columbia, Opdyke had been recruited by Runcorn to collect samples for paleomagnetic studies in the Grand Canyon. For such strenuous work, Runcorn had sought not only an undergraduate geology major but also a football player. Opdyke fit the bill. After receiving a Ph.D. from Newcastle, Opdyke worked on paleomagnetism in Australia and Africa. He arrived at Lamont just before the plate tectonic revolution, convinced that continental drift was a reality.

William Ludwig of Lamont also took continental drift seriously, as demonstrated in a seminar he gave in the early 1960s. He worked on the extensive set of marine geophysical data Lamont collected during the International Geophysical Year from the Falkland plateau and the Scotia

Sea in the far southern Atlantic. He was familiar with the proposed fit of the coastlines of Africa and South America by South African geologists, including evidence that older geologic formations matched on the two sides of the South Atlantic. Ludwig realized that the Falkland plateau, which he showed consists of continental crust that juts eastward from Argentina, fits neatly beneath the south coast of South Africa when the Atlantic Ocean is closed.

Deep Earthquakes and the Underthrusting of Lithosphere at Island Arcs

In 1965, my work on island arcs and our seminar on deep earthquakes led me to join Bryan Isacks, who received his Ph.D. from Columbia about the same time I did, to travel to Fiji and Tonga to study deep earthquakes beneath that region. Prior to my arrival, Bryan had set up two seismic stations in Fiji and one in Tonga. Bryan and I puzzled about the recordings of high-frequency seismic waves at the station in Tonga from intermediate- and deep-focus earthquakes between Tonga and Fiji—that is, those with depths of 45 to 430 miles (70 to 690 kilometers).

While in Fiji, I traveled by air and boat to the remote Tongan island of Eua, which is the closest island to the Tongan deep-sea trench. Using a portable seismograph to record earthquakes for a week, I obtained a seismogram from a large deep earthquake that contained very high-frequency seismic waves, even more spectacular than those at our existing Tongan station farther west.

A local Tongan helped me find that site for my portable instrument. He told me that the last time a foreigner had visited Eua, a very bright flash had occurred in the sky. It must have been from one of the large, high-altitude nuclear explosions detonated by the United States between Tonga and Hawaii. He did not say, but I think he believed I was there to record and witness another such flash.

In 1967, Oliver and Isacks concluded that the high-frequency waves I studied and others traveled up the inclined earthquake zone. They proposed that the dipping seismic zone was associated with lithosphere that had been underthrust beneath the Tongan arc to great depths. Lithosphere is the cold, strong outermost part of the earth. It typically consists of crust and uppermost mantle. It is underlain by the asthenosphere, a weak layer of low, long-term strength that is the "gliding layer" of plate tectonics. Lithosphere and asthenosphere are shown schematically in figure 1.2.

The terms *crust* and *mantle* are used more often in the earth sciences than *lithosphere* and *asthenosphere*. The crust and mantle of the earth are distinguishable by their chemistries. A sizable increase in density and the speed of seismic waves occurs where the crust meets the uppermost mantle at what geophysicists call the *Mohorovicic* (Moho or M) *discontinuity*. Mineral content also changes crossing that boundary. The earth's outermost layer on which we live is the crust. It is very thin beneath most oceans and much thicker beneath continents. Rocks within continents are as old as 4 billion years old, whereas those beneath the oceans are much younger, less than 200 million years old, about 5 percent of the age of the earth. The mantle is the largest component of the earth and extends about halfway to its center (figure 3.1). The core of the earth contains mainly iron. The outer core is liquid, and the inner core is solid.

Prior to the recordings in Tonga and Fiji, Isacks, Oliver, and I presumed that the presence of the young Tongan volcanoes implied that the downgoing seismic zone was hotter than other parts of the mantle and would greatly attenuate or impede the propagation of seismic waves. In fact, we observed just the opposite: those waves propagated very efficiently up the inclined seismic zone to the Tongan islands from deep earthquakes farther west. That finding was key for Isacks and Oliver's hypothesis in 1967 that oceanic lithosphere was underthrust at island arcs and for the development and testing of plate tectonic concepts (see Isacks, Oliver, and Sykes 1968).

I helped Isacks install more seismic stations on the two largest Fijian islands. Fortunately, I had read a great deal about Fijian customs before my arrival. One was that drinking a potent liquid called kava was ceremonial and that visitors must drink a cup down without hesitation. This knowledge was invaluable when Bryan and I camped near a small remote Fijian village while installing one station. We were invited to the village, offered chairs while others stood, and kava was produced to welcome us. From my nearly bald head, they took me to be the leader, not Bryan, and offered me the first cup of kava. I drank all of it, as expected. It had a slightly numbing effect on my throat. I had a few more cups as the evening and dinner progressed, but I did not feel especially inebriated.

Earlier in my work relocating earthquakes for the main down-going seismic zone beneath Tonga and Fiji, I discovered a separate group of very deep earthquakes farther west between Fiji and the New Hebrides (Vanuatu) island arc. In a short paper in 1964, I reported twenty earthquakes in the narrow depth range of 370 to 400 miles (600 to 650 kilometers). These earthquakes were not connected to the inclined seismic zones beneath either Tonga–Fiji or the New Hebrides (Vanuatu) island arcs. They likely occur in seafloor that was previously subducted beneath the Vitiaz trench to the northwest of the present hook at the northern end of the Tongan arc south of Samoa (figure 4.2). That trench is no longer seismically active today. Those deep earthquakes appear to lie near a major velocity and density change in the earth at a depth of about 400 miles (650 kilometers). It involves a rearrangement of the crystal structure of mantle minerals so that their density becomes greater below this depth.

5

SUBDUCTION, PLATE TECTONICS, AND THE NEW GLOBAL TECTONICS, 1967–1969

After the publication of papers on the seafloor's magnetic field in 1966 and on transform faulting, several of us began analyzing Lamont's huge databases related to seafloor spreading and continental drift. These searches led to fruitful interactions among various groups and individuals there. The magnetics group used an extensive data set to map and date magnetic anomalies for much of the seafloor of the Atlantic, Indian, and Pacific Oceans. Marine geophysicist Marcus Langseth examined measurements of high heat flow emanating from the oceanic crust.

Soon after the giant Alaskan earthquake in 1964, George Plafker of the USGS embarked by kayak on a field survey of its rupture zone. He found a consistent pattern of shoreline uplift and sinking inexplicable without invoking thrust faulting on a shallow-dipping fault in a huge region about 500-by-120 miles (800-by-200 kilometers). Frank Press and the graduate student David Jackson of MIT deduced that slippage in 1964 occurred on a nearly vertical fault that extended to great depth into the earth's mantle. They incorrectly picked the steep plane of the focal mechanism as the fault that ruptured.

In his epic paper "Tectonic Deformation Associated with the 1964 Alaska Earthquake" (1964), Plafker showed instead that the floor of the Pacific Ocean was being underthrust beneath southern Alaska along a *shallow-inclined plane*, ruling out the steep plane chosen by Press and Jackson. His identification of this major earth process, soon to be called *subduction*, was Plafker's major contribution to plate tectonics. He and

James Savage of the USGS went on to find similar results for the giant Chilean earthquake of 1960.

In 1967, Donald Tobin and I computed better locations of earthquakes and determined mechanisms of seismic events along the coastal region of western Canada and southeast Alaska. Our work, published in 1968, found that strike-slip faulting occurred well to the east of the 1964 rupture zone along large parts of the Pacific–North American plate boundary.

Subduction of the Lithosphere Along the Tonga Arc

Around the same time, Oliver and Isacks, two of my colleagues in seismology at Lamont, resurrected the concepts of *lithosphere*—the cold, strong outer part of the crust and uppermost mantle of the earth—that is underlain beginning at a depth of about 60 miles (100 kilometers) by warmer *asthenosphere*—material closer to the melting point of rocks and with low long-term strength (figure 1.2). The concepts of lithosphere and asthenosphere had been used more than 100 years earlier by geodesists to explain the equilibrium (isostatic balance) of large areas of differing elevation and crustal thickness. Beneath large areas, pressures at a depth of about 60 miles (100 kilometers) are equal, a condition called *isostatic balance*. These concepts of strength had been little used since Reginald Daly of Harvard described them in his book *Strength and Structure of the Earth* (1940).

Oliver and Isacks deduced that the inclined seismic zone in the Tonga island arc was caused by underthrusting of strong Pacific lithosphere beneath Tonga and Fiji to depths of 430 miles (690 kilometers). Earthquakes occur in that lithosphere because it is cold (recently underthrust) compared to the higher temperatures of the rest of the earth's mantle. Because the underthrust lithosphere is colder, its greater density contributes to its sinking into the earth's mantle. Down-going zones of lithosphere are abnormal in the sense that they have been underthrust at subduction zones. Deep earthquakes occur only in underthrust lithosphere, which makes up a very small percentage of the earth's upper mantle. Seismic waves from deep earthquakes propagating nearly upward to Fiji, however, lack high-frequency waves because they travel through normal, hotter upper mantle.

AGU Meeting of 1967 and the Birth of Plate Tectonics

The annual AGU meeting in Washington, D.C., in April 1967 was extremely important because papers presented there proved definitively the existence of seafloor spreading, transform faulting, subduction, and continental drift. It was at this meeting that most earth scientists first heard about those results. I spoke about my work on transform faults, including additional mechanisms from the Gulf of California and the Gulf of Aden. Oliver and Isacks presented their paper "Deep Earthquake Zones, Anomalous Structures in the Upper Mantle, and the Lithosphere" (1967).

Jason Morgan of Princeton extended the concept of seafloor spreading to the earth's spherical surface. He proposed that the earth's surface is made up of about twenty huge, cold, and rigid spherical shells, about 60 miles (100 kilometers) thick. He proposed that those shells constituting the lithosphere are bounded by Mid-Oceanic Ridges, where new surface is formed; deep-sea trenches and young fold mountains where surface area is being consumed (thrust back into the earth); and great faults (transform faults) where large horizontal displacements occur. The geophysical community soon called those spherical shells *plates*. And thus plate tectonics was born.

On a spherical surface, the motion of one plate with respect to another may be described by a rotation about what is called a *center of rotation* or *Euler pole*. Mathematician and physicist Leonhard Euler long ago described the rotation of two rigid spherical shells around one point on the surface of a sphere, at a center or pole of relative rotation. Transform faults along a plate boundary on a sphere form small circles around that pole of rotation. (Such poles should not be confused with the earth's axis of rotation, however.) This formalism can be expressed mathematically, but knowing the mathematical formula is not necessary for a general understanding of plate tectonics.

Morgan used existing data on magnetic anomalies at ridge crests and the orientations of recent transform faults along the Mid-Atlantic Ridge to compute a pole of rotation for the North American plate relative to the African plate for the past few million years. He found that the spreading of the Pacific-Antarctic Ridge in the southeastern Pacific, where the "magic magnetic profile" was obtained, showed the best agreement with his hypothesis. He calculated relative motion and poles of rotation for it and several other pairs of adjacent plates.

Morgan presented his work at this same spring AGU meeting in April 1967. Just before the meeting, he sent me an early draft. Although not well written, the paper nevertheless contained his basic ideas about what was soon called *plate tectonics*. I grasped the importance of his ideas. I don't think many people at the AGU meeting, however, appreciated what Morgan had to say.

Dan McKenzie of Cambridge University and Robert Parker of UC San Diego analyzed the motion of the Pacific and North American plates on a sphere using the slip directions from thrust earthquake mechanisms along that boundary. They submitted their work to *Nature*, and it was published very quickly, in December 1967. Morgan, who had presented his work at an earlier AGU meeting, submitted his work for publication in 1967. Upon revision, his famous paper "Rises, Trenches, Great Faults, and Crustal Blocks" was published in 1968.

Whereas Morgan's paper was globally inclusive, McKenzie and Parker's was not. Nevertheless, each paper outlined models for the relative movement of large, undeformed spherical shells—the tectonic plates—on the surface of the earth. In his major historical summary of earth mobilism, *The Continental Drift Controversy*, vol. 4: *Evolution Into Plate Tectonics* (2012), historian Henry Frankel describes the two independent discoveries and the correspondence between them in considerable detail, some of which was not known to me until 2012. Frankel interviewed many of the contributors to continental drift and plate tectonics. His work is unlikely to be surpassed because many of his sources have since died. The four volumes of *The Continental Drift Controversy* are an excellent and very thorough description of tectonic concepts that grew from continental drift a hundred years ago to paleomagnetism and plate tectonics.

At an AGU symposium for the fiftieth anniversary of plate tectonics in 2017, one organizer said that McKenzie, Parker, and Morgan discovered plate tectonics, as if nothing important had preceded their work. But seafloor spreading along Mid-Oceanic Ridges, transform faulting, and continental drift were well known before their work. Oliver and Isacks's paper published in 1967 was crucial in establishing that underthrusting of lithospheric sheets occurs at what soon were called *subduction zones*.

About the time of the spring AGU meeting in 1967, Xavier Le Pichon, who was then at Lamont and was aware of Morgan's results well before their publication in 1968, began using the extensive archive of seafloor data at

Lamont to extend Morgan's plate model to data from the entire globe. Le Pichon calculated the relative movements of a number of large crustal plates using data on rates of seafloor spreading and the directions of transform faults. He went on to use those inputs to calculate unknown directions and rates of plate motion at many subduction zones and regions of continental convergence, such as the Himalayas.

Le Pichon concluded that without substantial amounts of underthrusting, unrealistic and different amounts of earth expansion must have occurred around different vertical axes through the earth. He also extended his results further back in time to reconstruct a history of seafloor spreading during the entire Cenozoic Era, which is the past 65 million years of Earth's history. He submitted his paper "Sea-Floor Spreading and Continental Drift" in January 1968, and it was published that June.

The New Global Tectonics

After the spring AGU meeting in 1967, and once Oliver, Isacks, and I learned about the as yet unpublished results of Morgan's and Le Pichon's work, the three of us decided to write a comprehensive paper: "Seismology and the New Global Tectonics" (Isacks, Oliver, and Sykes 1968). It focused on earth mobility on a global scale. We pulled together information from many seismological studies plus new data of our own that related to seafloor spreading, continental drift, transform faulting, subduction, and motions of spherical plates of lithosphere.

Each of us made nearly equal contributions to the manuscript. We decided in 1968, near the end of preparing the paper, that we would draw straws to determine who would be listed as first, second, and third authors. Isacks drew first place, and I third. On the first page of our paper, a footnote states, "Order of authors determined by lot." Normally the person who makes the greatest contribution is listed as the first author, but this paper was an unusual case of equal contributions by three persons.

For our paper, Oliver drew a schematic cross-section of the outer 430 miles (690 kilometers) of the earth extending from the Tonga and Vanuatu (New Hebrides) subduction zones at the left, across the East Pacific Rise at the center to the Peru-Chile Trench and west coast of South America on the right (given as figure 1.2 in this book). We found that this figure, widely reproduced, helped readers to understand the key concepts of lithosphere, asthenosphere, and plate tectonics. It is carved on Jack Oliver's tombstone, which I presume Oliver stipulated in his will since his wife predeceased him.

An arc-to-arc transform fault at the left side of figure 1.2 connects the ends of the Tonga (T) and Vanuatu subduction zones, which dip in opposite directions. At the center of the figure, two ridge-to-ridge transform faults connect ridge crests of the East Pacific. The Pacific plate extends from the Tonga trench to the crest of the East Pacific Rise. The Nasca plate, which stretches from the East Pacific Rise to the west coast of South America, moves easterly (to the right) with respect to the Pacific plate. The Nasca plate underthrusts the South American plate at the right along the Peru–Chile subduction zone.

Isacks and I collected and analyzed many mechanism solutions of shallow earthquakes for several subduction zones. From those solutions, we plotted the directions one plate moved with respect to another (these directions are referred to as *slip vectors*). We very quickly found that these directions agreed quite well with Le Pichon's calculated motions of the lithosphere at subduction zones. Le Pichon used six major plates. He did not use some of the moderate-size plates that were recognized within the next two years. We found, nevertheless, that six plates were sufficient in 1968 for a first-generation analysis of plate motions.

The focal mechanisms of shallow earthquakes at island arcs strongly confirmed the hypothesis that subduction was one of the three major global modes of earth deformation. Our mechanisms put the cap on the theory of the underthrusting of lithosphere at subduction zones, one of the major conclusions of our paper in 1968. Our paper was considered by many to be the definitive description of large-scale earth deformation. The theory of the underthrusting of lithospheric plates at subduction zones removed problems related to proposed large amounts of earth expansion during a small fraction of the earth's age.

In our paper, we revisited the work I did in 1967 on transform faulting and normal faulting at ridge crests and in East Africa, including new mechanism solutions of earthquakes in the Gulf of California and the Gulf of Aden. We also added new mechanisms for branches of the Mid-Indian Ridge system, the Azores-Gibraltar Ridge, and the Queen Charlotte and Fairweather Faults off British Columbia and southeast Alaska. Slip vectors of all of those earthquakes agreed well with the directions of plate motions computed by Le Pichon.

Marine geophysicists had observed normal faulting beneath the deepest parts of trenches and their outer (seaward) walls, which seemed to contradict the notion of compression at subduction zones. We obtained data on normal-faulting mechanisms, where lithospheric plates were being bent downward just prior to their being underthrust beneath the inner (landward) walls of trenches. Those normal-faulting mechanisms represented deformation *within* the down-going plate, not slip between two plates. At the plate interface, we found thrusting and horizontal compression. Similarly, earthquakes deeper than about 40 miles (70 kilometers) occur *within* down-going lithosphere at subduction zones, not at the plate interface. Our findings made mechanical sense and resolved the apparent contradiction of normal faulting at subduction zones.

Another major conclusion of our paper was an answer to why deep earthquakes occur only in a few parts of the earth's upper mantle. We found they occur in cold, strong lithosphere that had been underthrust at subduction zones within the past 10 to 20 million years. Most of the earth's mantle deeper than about 40 miles (70 kilometers) is devoid of earthquakes.

Shallow earthquakes occur more widely and mostly along plate boundaries. They occur infrequently in the interiors of plates. Examples of the latter type of earthquakes, called *intraplate shocks*, include New Madrid in the central United States in 1811–1812 and Charleston, South Carolina, in 1886. Intraplate earthquakes are described in later chapters.

Most of the world's largest earthquakes from 1900 to 2018 occurred at shallow depths along subduction zones. Examples include the giant shocks in Chile in 1960, Alaska in 1964, Sumatra in 2004, and Honshu, Japan, in 2011. Their magnitudes were all greater than 9.0, and their rupture areas along the interface between two plates were very large. Their smaller dimension down the plate interface, being more than 150 miles (240 kilometers), resulted from the rupture plane dipping (plunging) very shallowly, only about 10 to 30 degrees from horizontal. That shallow dip leads to larger rupture zones and magnitudes than those for transform faults, which are nearly vertical and dip about 90 degrees. The measured seismic magnitudes of the largest known earthquakes along transform faults, such as the California shock in 1906, are about magnitude 8.0 and are not giant shocks like those along subduction zones.

The largest normal-faulting events along slow-spreading ridge crests such as the Mid-Atlantic Ridge are about magnitude 6.5, and those along fast-spreading ridges such as the East Pacific Rise are no larger than about magnitude 3.0. These smaller magnitudes result from temperatures being higher at a given depth beneath ridges, especially fast-spreading ones. Hence, their rupture depths are very small.

Vertical cross-sections of precisely relocated earthquakes at subduction zones are illustrated for Tonga and the northern Marianas in figures 4.1 and 5.1. The down-going seismic zones are remarkably narrow. The distribution of shocks beneath the northern Marianas is vertical and remarkably simple between depths of 100 and 430 miles (150 and 690 kilometers).

The paper Isacks, Oliver, and I published in 1968 helps to explain why earthquakes at some subduction zones such as the Aleutians extend to depths of only 150 to 200 miles (250 to 300 kilometers), whereas others such as those in Tonga and the Marianas extend deeper, to nearly 430 miles (690

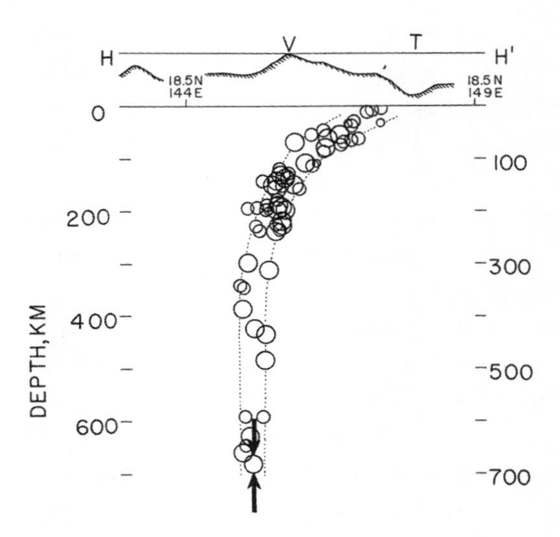

FIGURE 5.1 Vertical cross-section of relocated earthquakes beneath the northern Mariana subduction zone. Bold converging vertical arrows at a depth of 430 miles (690 kilometers) denote the direction of maximum compression as determined from a focal mechanism of a very deep earthquake.

Source: Katsumata and Sykes 1969.

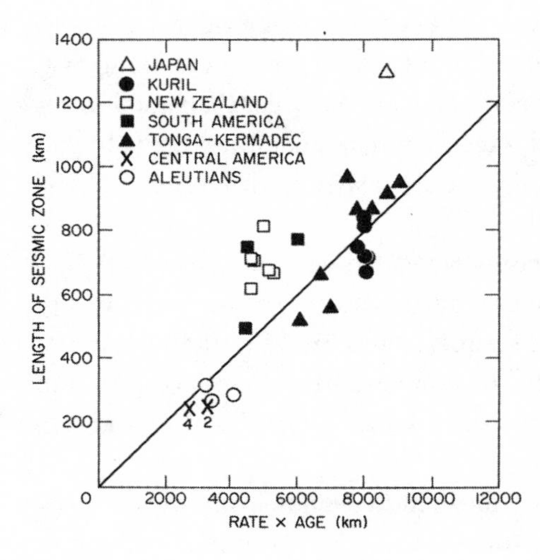

FIGURE 5.2 Lengths of down-going seismic zones for individual subduction zones on the vertical axis as a function of the rate of underthrusting multiplied by the age of the plate as it enters a subduction zone.

Source: Unpublished figure by the author, 1975.

kilometers). The measured lengths along inclined seismic zones are plotted on the vertical axis of figure 5.2. The depth at which the down-going lithosphere reaches temperatures where the earthquake process is halted (about 1,100°F [600°C]) is a function not only of the rate of underthrusting but also of the thickness and age of the plate being subducted. Because older plates are cooler, rate multiplied by plate age is plotted on the horizontal axis of figure 5.2, which shows a good correlation between the vertical and horizontal axes.

Subduction zones with high rates of underthrusting of old Pacific lithosphere, such as in Tonga and Japan, have long down-going seismic zones that extend to depths of nearly 430 miles (690 kilometers). Younger lithosphere is being subducted at slower rates beneath the Aleutians and Central America; hence, the maximum depths of earthquakes in these regions are about 150 to 200 miles (250 to 300 kilometers).

For our paper, we mapped additional paths where high-frequency P and S waves propagate efficiently. Older parts of the oceanic and continental lithosphere exhibit efficient propagation, as between earthquakes in the Greater Antilles and seismic stations in Bermuda and those along the east coast of the United States. Lower temperatures at depth are needed for the lithosphere to include the uppermost mantle and to allow those high-frequency waves to propagate.

Plate tectonic theory provided a new understanding of why most earthquakes, volcanoes, and young mountain belts occur where they do. Volcanoes at subduction zones typically have a steep slope like that of Mount Fuji in Japan and Mount Rainer in Washington State. Magmas injected at Mid-Oceanic Ridges or in Hawaii, in contrast, are typically less viscous and form volcanoes with gentle slopes. Most large tsunamis are generated by the sudden uplift or subsidence of the seafloor during a large earthquake at a subduction zone.

James Brune of Lamont devised a method to sum the contributions from many earthquakes along a given plate boundary and thence to calculate a quantity called *seismic moment* from very long-period seismic waves. He used those summations to calculate the average rate of plate motion from earthquakes. The largest earthquakes along a plate boundary contribute most to that sum. In our paper, Isacks, Oliver, and I used Brune's method to calculate the amount of relative plate motion at the Tongan, Japanese, and Aleutian subduction zones from shallow earthquakes. Our calculations agreed with Le Pichon's calculations of plate motions at those island arcs to within a factor of two, which we considered to be a considerable achievement at the time.

Lamont graduate student Peter Molnar and I analyzed the plate tectonic framework of the Caribbean, Mexico, Central American, and northwestern South America. For our paper "Tectonics of the Caribbean and Middle America Regions from Focal Mechanisms and Seismicity" (Molnar and Sykes 1969), we used about 600 relocated earthquakes and seventy new earthquake mechanism solutions that we obtained. Our work showed that the Caribbean was a separate plate sandwiched between the North American and South American plates. We gave the name "Cocos plate" to a new plate to the south of Mexico and Central America, naming it for the Cocos Ridge, along which Cocos Island rises above sea level.

We also named another new plate, the Nasca plate, which is bordered on the north by the Galapagos Rift, on the west by the East Pacific Rise, on the south by the West Chile Ridge, and on the east by the subduction zone along the west coast of South America. We named it for the aseismic Nasca Ridge on that plate and for the ancient Nasca culture. We also confirmed that subduction occurs along the Lesser Antillean island arc in the eastern Caribbean and along the southwestern coasts of Mexico and Central America.

Importance of Plate Tectonics

Soon after 1968, most geophysicists in North America accepted the plate tectonic model. Few people questioned the basic concepts after 1969. It was a stunning scientific revolution. Our understanding of our planet was rapidly and radically transformed. This scientific revolution is recognized as being as important to understanding the earth as the development of quantum mechanics was important to physics and evolution and DNA were to biology and medicine.

I presented an invited paper, "Frontiers in Geophysics: The New Global Tectonics," at the fiftieth anniversary meeting of the AGU in April 1969. Plate tectonics is now taught in many elementary schools. The acceptance of it was slow for many in the petroleum industry, but today it is an integral part of exploration for oil, gas, and coal.

I add here an aside about my employment during the development of plate tectonics. I worked at Lamont for the U.S. Department of Commerce for two years during the preparation of the papers published in 1967 and 1968 (Sykes 1967; Isacks, Oliver, and Sykes 1968). I was concurrently an adjunct assistant professor at Columbia. James Brune had held that position prior to me, Bryan Isacks after me. A bonus of having that position at Columbia was that Lamont received film chips for all of the WWSSN and Canadian seismograph stations from the U.S. government, which helped to support our work.

When George Sutton left Lamont, Oliver concluded that Columbia needed to have another tenure-track professor in seismology. He approached the president of Columbia and the Department of Geology about giving that position to me. As a result, I left federal employment and became an assistant professor of geology at Columbia in August 1968. The following year I was promoted to associate professor, and in 1978 I became the Higgins Professor of Geology (now Earth and Environmental Sciences).

6

EARTHQUAKES IN THE CARIBBEAN AND ALASKA

Before I go on in the next chapter to describe work that started on long-term earthquake prediction, it is necessary to cover work I did in the Caribbean and Alaska, mostly in the mid-1960s.

I redetermined the locations and depths of a large number of earthquakes along the borders of the Caribbean Sea. Over the next 50 years, I returned several times to work on earthquakes and the tectonics of the Caribbean and Central America. My improved locations of earthquakes in 1965 more clearly delineated the down-going seismic zone beneath the Lesser Antilles and Puerto Rico. The maximum depths of earthquakes there, unlike the depths of earthquakes in Fiji and the northwestern Pacific, did not exceed 150 to 200 miles (250–300 kilometers).

In 1970, I drove up a very steep, narrow road to a point near the summit of the Grande Soufrière volcano in Guadeloupe in the eastern Caribbean after spending a week's vacation on the island. The smell of sulfur was strong, and it was foggy. I had the place to myself. I also visited the base of Mont Pelée volcano in Martinique in 1973 and toured the town of Saint Pierre, which was destroyed by a glowing avalanche during the volcano's major eruption in 1902. It killed 30,000 people. The only survivor was fortunate that he had been held in a deep jail.

In 2002, my wife, Kathy, and I spent a week in Tortola, British Virgin Islands. We traveled for a day to Virgin Gorda, with its extensive outcrops of granites and beautiful beaches. The granites probably were displaced eastward from the Greater Antilles as the Caribbean plate moved easterly from North America.

Alaska

In 1966, Donald Tobin and I relocated about 300 earthquakes beneath mainland Alaska, the Alaska Peninsula, and adjacent offshore areas in the 10-year period before the great Alaskan earthquake of 1964. Don was a postdoc who had completed a Ph.D. at Columbia in structural geology. I raised funds for his salary and our work. We found that the rupture zone of 1964 exhibited only minor earthquake activity in the preceding 10 years.

That work was inspired by my previous fieldwork during the summer of 1964, when Tosimato Matsumoto, a research scientist at Lamont, and I recorded aftershocks of the giant Alaskan event in 1964. I felt my first earthquake during that trip, a nearby aftershock of about magnitude 5.5. I had just slipped into bed about 2:00 a.m., and Tosi had gone to the outdoor toilet. He came running out, pulling up his pants, and saying, "Did you feel it?" I had indeed. I knew it was an earthquake right away but not large enough to do any damage to our log cabin. Wooden structures usually fair very well in strong earthquakes.

During the summer of 1967, I worked in central Alaska along the Denali fault with Robert Page of Lamont and the graduate students Peter Molnar and Ramesh Chander, trying to measure stress buildup along a branch of that fault. Walking 10 to 15 miles (16 to 25 kilometers) on some days through swampy ground along the fault zone was one of the hardest physical tasks I have ever done. Many times we stepped into holes in the swamp produced by moose hooves and sank into water deeper than our boots. The mosquitoes were huge. We had to wear hats with netting over our faces and necks when we were in the field. The days were long in June, but it was cold—in the upper 30s to low 40s F (about 3° to 7°C).

We drove copper-coated steel rods into the earth about 0.3 miles (0.5 kilometers) apart on both sides of the fault and surveyed their locations. Page came back in later years to see if the rods' locations had been altered by stresses built up along the fault. They had not, which meant we had missed the main fault. We drove the rods to "refusal" using a sledgehammer—very hard work. We had to use a machete and axes to clear lines of sight between the rods so we could survey them.

Jack Oliver, Chander's Ph.D. adviser, had sent him to Alaska with us to get field experience. Except for a trip to the Adirondacks, Chander had never been outside either New Delhi or New York City. He had no idea how to use a machete. I had to forbid him from using it after seeing him try to cut brush because I was afraid he would cut his leg. We were in a very wild part of Denali (McKinley) National Park. A helicopter would not have been available to rescue an injured person.

On the way to our field site, we had to drive about 150 miles (240 kilometers) along the unpaved Denali highway. We stopped at one of the food-and-gas stores spaced about every 40 miles (65 kilometers) along the highway. As we entered one store, Chander was aghast to see a huge stuffed Kodiak bear in the two-story entrance that was mounted vertically to a height of about 15 feet (4.6 meters). His only happy day in the field was the day we left.

About 10 miles (15 kilometers) east of our field site was a prominent fault scarp. A scarp is a sudden change in elevation from one to many feet and is formed along a fault when movements during past earthquakes offset topography. I thought the most recent movement at that scarp could not have been more than about 100 years earlier because winter weather is fierce there and the rate of erosion is high.

A few years later I relocated a number of moderate-to-large earthquakes that had occurred in mainland Alaska between 1904 and 1953. Surprisingly, enough data existed to make a fairly precise relocation of a great earthquake that occurred in central Alaska in 1904. It was not on the major Denali fault, as I first suspected, but well to the northwest of Fairbanks in central Alaska. Hence, the last earthquake on the fault scarp to which we walked likely occurred before 1904. A large forerunning shock to a great earthquake of magnitude 7.9 in 2002 ruptured near that scarp and to the east of it along the Denali fault (figures 6.1 and 6.2).

In 1967, we saw many signs of grizzly bears during our fieldwork. We carried whistles to warn them of our approach. Alaskans told us we were foolish not to carry powerful rifles, but we didn't because they were forbidden in the national park where we worked. We walked along another part of

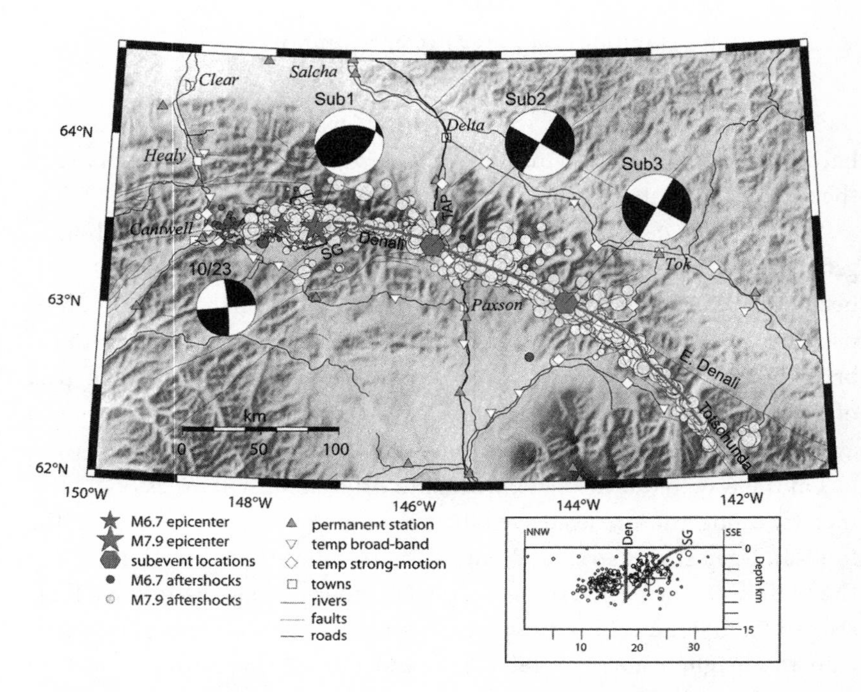

FIGURE 6.1 Circles denote aftershocks of a great earthquake along the Denali Fault on November 3, 2002. Its magnitude, 7.9, was comparable to that of San Francisco earthquake of 1906. See plate 2.

Source: Eberhart-Phillips et al. 2003, with permission of the American Association for the Advancement of Science.

FIGURE 6.2 Offset of a highway in central Alaska during the great earthquake of 2002 along the Denali fault. The Trans-Alaskan Pipeline crosses the fault near this location. Robert Page Jr. of the USGS insisted that the pipeline be built to accommodate such an offset, which it was. See plate 3.

Source: Photograph courtesy of Lloyd Cluff.

FIGURE 6.3 The author taking a core of a tree along the Denali fault in 1967. See plate 4.

Source: Photograph courtesy of Robert Page Jr.

the fault with a park ranger. We saw mountain sheep about 1,500 feet (0.5 kilometers) across a valley. We stared at one another. The ranger said they likely had never seen people before. We took cores from trees along the fault to try and date past large earthquakes using changes in the annual growth of tree rings (figure 6.3), but we were not successful.

7

LONG-TERM EARTHQUAKE PREDICTION, SEISMIC GAPS

Alaska, Mexico, and South America

Although a number of my geophysical colleagues believe earthquake prediction is not possible, I have devoted a considerable amount of time since the late 1960s to various aspects of prediction. My colleagues and I have worked on seismically active plate boundaries in Alaska, the Aleutians, Mexico, Colombia, Peru, and Chile. I focus here on long-term prediction (several decades), less on intermediate-term prediction (months to a decade), and not at all on short-term predictions (days to months).

Seismologists are often asked, "When and where will the next Big One occur?" My short answer is, "We don't know." Nevertheless, my longer answer is: "Prediction is still in its infancy, but we can do better with longer-term earthquake prediction for certain areas where intensive monitoring and study have been undertaken." Continued monitoring of several geophysical and geochemical effects is essential. This monitoring needs to be coupled with a better understanding of the physics of earthquakes, better engineering of structures, and ways to reduce damage and loss of life in earthquakes.

Terms Used to Describe Earthquakes and Tectonic Processes

Let me provide some more basic information about earthquakes. Stresses (pressures) build up in the earth over decades to hundreds of years near plate boundaries. The energy that was slowly accumulated is then suddenly released at the time of a large shock. That energy goes into heating rocks along faults and usually less into generating seismic waves. During an earthquake, displacement (slip) takes place along a fault either at the surface or at depth in the earth. Slippage can be either horizontal or vertical. I focus here on displacements that occurred during earthquakes of the past few thousand years. I call the structures on which they occur *active faults* because they have the potential to be the sites of future earthquakes.

The work I describe here concentrates on large, great, and giant earthquakes (table 7.1) along active subduction zones and transform faults. Relatively little of the energy released globally in earthquakes occurs along the Mid-Oceanic Ridges. The three main types of faults—thrust, normal, and strike-slip—are described in chapter 1. All giant and nearly all great earthquakes take place along thrust faults at convergent plate boundaries, such as subduction zones and regions of recent compressive mountain building, where two plates meet and one is thrust beneath the other, as in Japan, Alaska, South America, and the Himalayas.

Strike-slip faulting occurs along a nearly vertical plane in the earth, with displacement in a horizontal direction. Strike-slip faults at plate boundaries, such as the San Andreas Fault of California, are also transform faults. Nearly all of the largest earthquakes along transform faults are smaller than magnitude 8.0, whereas giant shocks at subduction zones have been as large as magnitude 9.6.

TABLE 7.1 Descriptions of Sizes of Earthquakes and Their Magnitudes

Giant earthquakes	Shocks of seismic magnitude greater than 8.5
Great shocks	Shocks of magnitude 7.75 to 8.5
Major earthquakes	Shocks of magnitude 7.0 to 7.75
Large earthquakes	Shocks of magnitude 7.0 and greater
Moderate-size earthquakes	Shocks of magnitude 5.0 to 7.0
Small shocks	Shocks of magnitude 3.0 to 5.0
Microearthquakes	Shocks of magnitude less than 3.0

The *hypocenter* is the place at depth in the earth where rupture initiates in an earthquake. Its projection onto the surface of the earth is called the *epicenter* (figure 3.1). Rupture in a large earthquake propagates outward along a fault in two dimensions from the initiation of fast slip at the hypocenter for tens to hundreds of miles. Shallow earthquakes involve thrust, normal, and strike-slip faulting and occur where temperatures in the earth are low enough that stresses can build up slowly. Depths in the earth where *shallow earthquakes* occur extend from the surface to about 30 to 45 miles (50 to 70 kilometers) deep. Damage is typically greater for shallow events than for deeper earthquakes of the same magnitude.

Two scales are used to describe the sizes of earthquakes. The older, called the *Modified Mercalli intensity scale*, uses Roman numerals to describe varying amounts of damage or perceived shaking. Intensities and magnitudes are often confused in press reports. I use magnitude scales to discuss earthquake prediction and seismic measurements of nuclear tests. Some older historic earthquakes in New York and elsewhere, however, can be analyzed only via reports of their maximum intensities in newspapers because seismic records and magnitudes do not exist for them.

Magnitudes have been used to describe the size of large earthquakes going back to the development of seismic instruments in the 1890s. *Magnitude scales*, which were invented in the 1930s, involve measuring the amplitudes or sizes of seismic waves on earthquake records called *seismograms*. Although seismologists use many different magnitude scales, I am concerned in this chapter and later chapters with one called the *moment magnitude scale* (Mw), which utilizes measurements of very long-period (very low-frequency) seismic waves. This scale captures the full extent of rupture and the size of great and giant earthquakes.

An increase in magnitude by one unit—for example, from 7.0 to 8.0—corresponds to about a thirtyfold increase in energy release. The energy release in the largest earthquake recorded since the invention of seismic instruments in the 1890s occurred during the giant earthquake in Chile of magnitude 9.6 in 1960. It ruptured about 550 miles (900 kilometers) of the plate boundary along the west coast of South America. The magnitude of the recent Japanese earthquake in 2011, the largest historical event in that country, was 9.0. By contrast, the largest underground nuclear explosion had a magnitude of about 6.9.

Giant earthquakes release so much energy because their rupture extends 125 miles (200 kilometers) or more down their very shallowly inclined fault planes at subduction zones before temperatures in the earth prohibit the buildup of stress. In contrast, the largest earthquakes along near-vertical strike-slip faults extend only as deep as about 10 to 15 miles (15 to 25 kilometers). The rupture zones of earthquakes at Mid-Oceanic Ridges are even more limited in size because high temperatures are reached at very shallow depths.

Earthquake Prediction and My Involvement

Here are the things that a good earthquake prediction should include:

1. An estimate of the earthquake's time of occurrence
2. The earthquake's approximate location and depth
3. The earthquake's size—that is, its seismic magnitude
4. An estimate of the probability that the event will occur
5. Uncertainties in parameters 1–4
6. A discussion of the scientific bases of the prediction
7. The time period for which the prediction is in effect and a stop date

Forecasts of the approximate amounts of damage, injuries, and loss of life also might be made in conjunction with an earthquake prediction. Accurately estimating the time of occurrence of a future earthquake is usually the most difficult aspect of a prediction.

My first interest in predicting earthquakes occurred in the mid- to late 1960s after my former thesis adviser, Jack Oliver, returned to Lamont from the first United States–Japan Conference on Earthquake Prediction, held in Japan. He was excited about the results of the meeting. I was then a new Ph.D. student working on plate tectonics. Oliver was instrumental in organizing and running a second conference hosted by Lamont, which I attended and helped run. I discovered then that a number of leading Japanese seismologists had a serious interest in predicting earthquakes. Five joint conferences, which I attended, were held alternately in Japan and the United States every few years.

Prior to 1968, few geophysicists in either the United States or Japan worked on the science of predicting earthquakes. I, too, had not taken earthquake prediction seriously enough to work on it. Likewise, during the two decades after World War II seismologists in Japan concentrated almost exclusively on studying either seismic surface waves or the physics of earthquakes. Little had been done on prediction in Japan since Akitune Imamura's study of what he described as unusual ground deformation in the days before the great earthquake along the Nankai subduction zone of Japan in 1944. Imamura did similar work on the very damaging Kanto earthquake in 1923, which ruptured the plate boundary just to the south of Tokyo and caused great damage and loss of life.

Believing earthquake prediction was not possible, Father James Macelwane, a Jesuit at St. Louis University and a seismologist, made very negative remarks about working on it. Charles Richter of Cal Tech was famously quoted as saying that earthquake prediction is for fools and charlatans. In the past 20 years, seismologist Robert Geller of Tokyo University has criticized work on prediction in general and prediction in Japan in particular, claiming that those making predictions merely seek publicity.

Soon after James Chadwick discovered the neutron in 1932, Ernest Rutherford, the father of the nuclear structure of the atom, said, "Anyone who looked for a source of power in the transformation of the atoms was talking moonshine." He anticipated neither the development of nuclear weapons nor the role of neutrons in their design. Rutherford died too early to see Leó Szilárd's idea of controlled nuclear chain reactions become a reality. Earthquake prediction thus should likewise not be written off as impossible.

Following my work on plate tectonics in the 1960s, I decided to apply my knowledge of very active plate boundaries to long-term earthquake prediction with a time scale of one to a few decades. That time scale was not very appealing to many seismologists, who said the public wanted accurate short-term predictions—that is, warnings of hours to weeks for very specific places. Geophysicists need to state clearly what we can and cannot deliver in predicting earthquakes either now or perhaps even in the next few decades.

Seismic Gaps and Long-Term Prediction

Knowledge of plate tectonics became useful in my work on long-term prediction. Starting in 1968, I examined long-term forecasts for what are called *seismic gaps*, segments of active plate boundaries that have not experienced large, great, or giant earthquakes for a long time. Seismic gaps may be considered to be rough qualitative estimates of the time remaining before a subsequent large shock will occur.

Some of my colleagues at Lamont and I decided to work on *great* earthquakes, those larger than magnitude 7.75, which account for most of the total energy released in earthquakes. Small shocks, although numerous, account for only a small fraction of the total energy release and do not act as a safety valve to prevent or delay the occurrence of large to giant earthquakes. Plate boundaries that had ruptured in large shocks fairly recently would not soon be the sites of earthquakes of similar size. We reasoned that stresses must be reestablished slowly by plate motion to about the level that existed prior to the most recent large shock. Based on our understanding of plate tectonics, we determined that a large earthquake does not "strike twice" at the same place within a time period that is short compared to the time needed to reestablish stresses.

Understandably, parts of plate boundaries that had not been the sites of large earthquakes for decades to hundreds of years were more likely to be the locations of large shocks during the next several decades. The seismic-gap hypothesis goes back to the work of the American geologist G. K. Gilbert in the late nineteenth century and to H. F. Reid's analysis in 1908 of the decades of slow stress buildup between 1857 and 1906 and its sudden release during the San Francisco earthquake of 1906.

In 1965 and 1968–1969, respectively, S. A. Fedotov of the Soviet Union and then Kiyoo Mogi of Japan used aftershocks to map the dimensions of rupture zones of past great earthquakes in the northwestern Pacific seaward of Kamchatka, the Kuril Islands, and Japan. They demonstrated that great earthquakes filled zones that had not been the sites of large to great shocks during the previous 30 to 100 years. My contribution was to put the work of Fedotov and Mogi into a plate tectonic framework.

Over the next decade, my Lamont colleagues and I set out to identify major seismic gaps for many active plate boundaries. Fedotov and Mogi as well as John Kelleher, Oliver, and I of the Lamont group had previously identified eight seismic gaps that subsequently became the sites of large to great earthquakes during the early 1970s. In the next section, I describe several of the plate boundaries and seismic gaps the Lamont group studied in the 1970s and 1980s.

Seismic Gaps in Southern Alaska, the Aleutians, and Offshore British Columbia

In 1968, Donald Tobin, a postdoctoral student at Lamont, and I published relocations and mechanisms of moderate to large earthquakes between northern California and south-central Alaska. We stated that parts of the Pacific–North American plate boundary in our study area that had not been the sites of major to great earthquakes for many decades were more likely to be sites of future large events. We called those segments of the plate boundary *gaps in seismicity.*

We proposed that a gap existed between the northwestern end of the rupture zone of the great Queen Charlotte Islands earthquake of 1949 and the southeastern end of the rupture zone in 1958 (figure 7.1). Half of this gap soon ruptured in a major shock of magnitude 7.6 offshore of Sitka, Alaska, in 1972. All or most of the remaining gap off southeastern Alaska broke in an earthquake of magnitude 7.5 (figure 7.2) in 2013. The gap in the westernmost Aleutians in figure 7.1 ruptured in a magnitude 7.7 shock in July 2017.

An unusual earthquake involving a combination of thrust and strike-slip faulting of magnitude 7.8 broke offshore of the Queen Charlotte Islands (Haida Gwaii) in 2012. The relative plate motion in that area contains a component of underthrusting as well as strike-slip motion. The latter was the dominant mode of slip in the great earthquake of 1949. Its rupture zone has

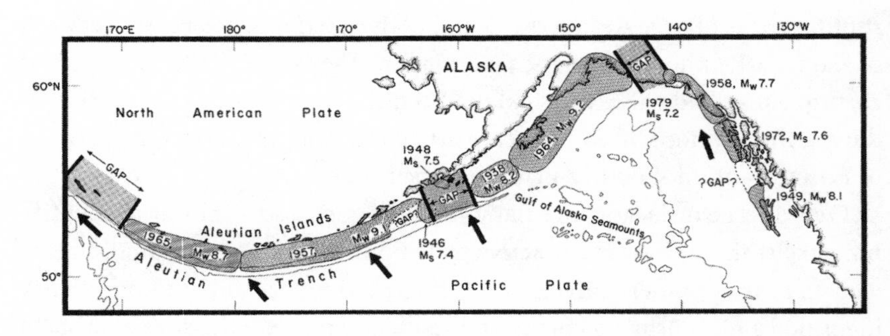

FIGURE 7.1 Seismic gaps for large earthquakes along the North American–Pacific plate boundary in the Aleutians, southern Alaska, southeast Alaska, and British Columbia.

Source: Unpublished figure by the author, 1979.

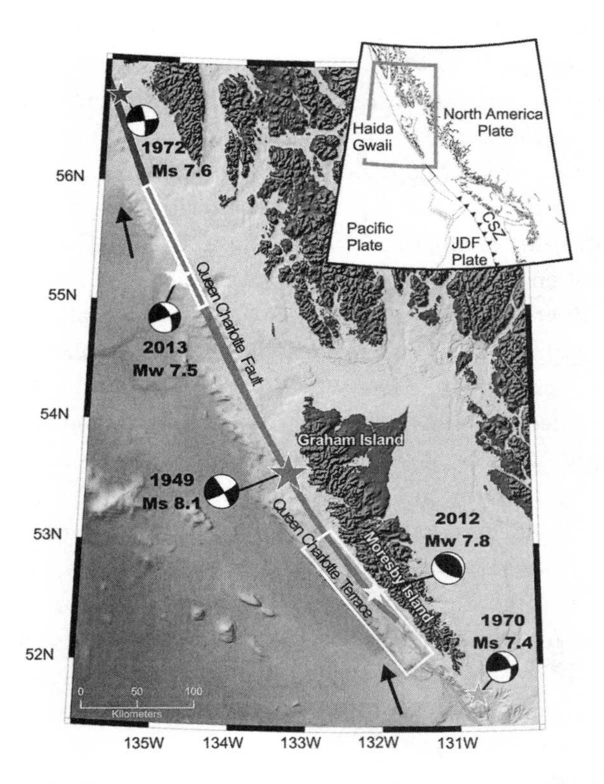

FIGURE 7.2 Map of Haida Gwaii, British Columbia, and Craig, Alaska, earthquakes of magnitudes 7.8 and 7.5 in 2012 and 2013. The latter event filled a seismic gap south of the earthquake rupture zone in Alaska in 1972. See plate 5.

Source: James et al. 2015.

not broken again and remains a seismic gap for a great future strike-slip event as of late 2018.

In 1971, I published a long paper, "Aftershock Zones of Great Earthquakes, Seismicity Gaps, and Earthquake Prediction for Alaska and the Aleutians." I examined the very active North American–Pacific plate boundary in southern Alaska and the Aleutians—that is, the region to the west of the region Tobin and I studied in 1968. Figure 7.1 shows my assessment in 1979 of seismic gaps along that plate boundary between offshore British Columbia and the westernmost Aleutians. Much of the subduction zone in southern Alaska and the Aleutians ruptured in great and giant earthquakes between 1938 and 1965. Three of those rupture zones were huge, more than 500 miles (800 kilometers) long.

Because plate boundaries transform to other tectonic features—subduction zones, transform faults, and spreading centers—they do not just end. Hence, seismic gaps along plate boundaries must either rupture in future large earthquakes, slip episodically without large shocks in what are called *slow-slip events*, or slip continuously. In the 1970s, we lacked crucial information to ascertain which gaps belonged to which of those three categories. But that information is now becoming available from studies of prehistoric earthquakes, Geodetic Positioning System (GPS) data, measurements of stress buildup and release, deep-seismic soundings of subduction zones, and detection of slow-slip events. Very long-period seismic magnitudes, Mw, which became available in 1976, led to a more accurate determinations of the sizes of great and giant earthquakes.

In 1971, I concluded that the seismic gap between the rupture zones of the southern and southeastern Alaska earthquakes in 1964 and 1958 (figure 7.1) had not broken in a great earthquake since 1900. A small part of that zone, now called the Yakataga seismic gap, ruptured in a moderate-size shock of only magnitude 7.2 in 1979. Hence, a considerable part of it still remains a seismic gap as of late 2018.

In a study of seismic gaps published in 1973, “Possible Criteria for Predicting Earthquake Locations and Their Application to Major Plate Boundaries of the Pacific and the Caribbean,” Kelleher, Oliver, and I proposed that the plate boundary between the great shocks of 1938 and 1957, which we called the “Shumagin Islands seismic gap” (between 160°W and 163°W in figure 7.1), had not ruptured for many decades. Some of my colleagues at Lamont showed later that about half of it broke in 1917 in a large but not a great earthquake. Geodetic measurements since 1975 indicate that stress buildup in much of the Shumagin region is anomalously low and that parts of it may well move slowly without large earthquakes. Although the Shumagin gap may not break by itself in great shocks, it may have ruptured in 1788 in conjunction with a giant earthquake along the adjacent plate boundary to the northeast.

As of late 2018, the rupture zones of the southeast Alaska, southern Alaska, and Alaska Peninsula shocks in 1958, 1964, and 1938 have not reruptured in events larger than magnitude 7.1. This finding is in accord with the seismic-gap hypothesis that large to great earthquakes do not recur until stress is slowly built up by plate motion. The rupture zone that broke in 1938 may now have built up much of the stress that was released more than 80 years ago. I regard it as more likely to break in a great earthquake during the next few decades than the rupture zones that broke in 1958 and 1964.

A small part of the rupture zone of the giant shock of magnitude 8.6 in the western Aleutians in 1965 broke in a major event of magnitude 7.7 in November 2003. Much of the rest of the rupture zone of 1965, however, has not broken again in earthquakes of significant size. Hence, most of it remains a seismic gap. The western third of the rupture zone of the giant Aleutian earthquake of magnitude 8.6 in 1957 reruptured in two events, each of magnitude 7.9 in 1986 and 1996. The rest of the rupture zone that broke in 1957 remains a seismic gap.

Each of the long rupture zones of 1957 and 1965 had broken in a series of large but not giant shocks in the early part of the twentieth century and may have begun to do so again starting in 1986 and 2003. A common feature of the rupture zones of the giant shocks of 1957 and 1965 is that they break at different times in a series of several large to great shocks rather than in a single giant earthquake. In subsequent work, Kelleher, Oliver, and I found that giant shocks at other subduction zones often rerupture in a similar way.

Global Studies of Seismic Gaps

In 1973, Kelleher, Oliver, and I expanded the Lamont studies of seismic gaps to include the active plate boundaries along the coasts of Central and South America, the Kuril Islands, Kamchatka, Japan, the Caribbean, and the San Andreas Fault of California. We paid particular attention to seismic gaps that appeared to be nearing the time of the next great event based on the time intervals between previous great historic earthquakes. Some other seismic gaps were difficult to evaluate because it was not known whether they had ruptured previously in large shocks.

Mexican Subduction Zone

In 1973, Kelleher, Oliver, and I found that many seismic gaps for large historic events (figure 7.3) off the Pacific coasts of Mexico and Central America were only of moderate length—60 to 125 miles (100 to 200 kilometers)—in contrast to the very long rupture zones of great-to-giant earthquakes in Alaska and the Aleutians (figure 7.1). Hence, accurate mapping of the locations of short seismic gaps off Mexico and Central America is more difficult. We found that the repeat times of six major but not great earthquakes off the coasts of Mexico and Central America were about 30 to 40 years, shorter than the repeat times of many great-to-giant shocks elsewhere.

One of the larger seismic gaps identified in the paper we published in 1973, which had previously broken in 1911, was the site of the great Michoacán earthquake of magnitude 8.0 on September 19, 1985, off the west coast of Mexico (westernmost gap in figure 7.3). That segment has a somewhat longer repeat time and generates larger earthquakes than most other parts of that plate boundary. The 1985 event, the most destructive earthquake in the history of Mexico, caused extensive damage and loss of life in Mexico City.

A shock of magnitude 7.5 two days later extended the rupture zone farther to the southeast. Such an expansion within hours to weeks is not uncommon. For example, a shock of magnitude 7.9 occurred about half an hour after the giant Japanese earthquake of March 11, 2011. It extended the

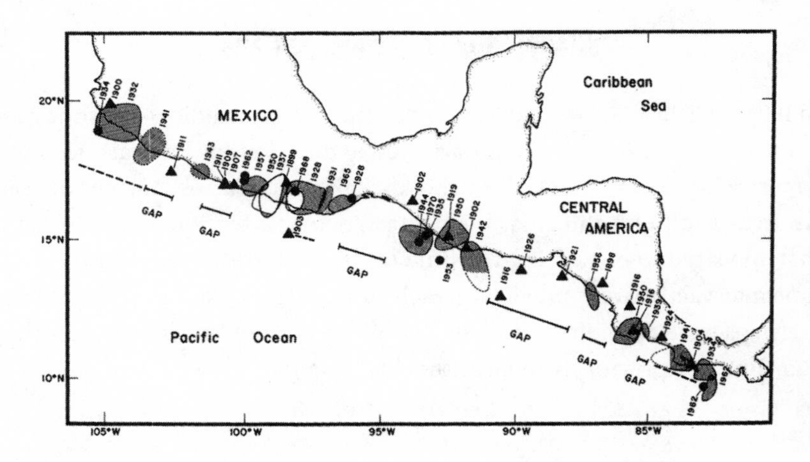

FIGURE 7.3 Seismic gaps for large earthquakes as of 1972 along the west coast of Mexico and Central America. The Michoacán earthquakes of 1985 filled the westernmost gap, which had ruptured previously in 1911.

Source: Kelleher, Sykes, and Oliver 1973.

rupture zone of the giant earthquake about 100 miles (150 kilometers) farther to the southwest. Because the magnitude 7.9 event occurred closer to Tokyo, it, not the preceding giant shock, caused much of the damage to populated and industrialized prefectures just to the east of Tokyo. Preceding larger earthquakes certainly triggered both this subsequent shock in Japan and the second shock in Mexico in 1985.

Mexican and U.S. seismologists are making special efforts to monitor the Guerrero seismic gap along the southern coast of Mexico, near longitude 101°W to the west of the shock in 1957 and to the east of the shock in 1943, as shown in figure 7.3. The Guerrero gap ruptured in a series of large events early in the twentieth century. Mexican geophysicists recently concluded from GPS data that much of that gap moves in large slow-slip events without strong shaking. Hence, the risks to Mexico City are likely not as great as was thought just a few years ago.

Because repeat times are relatively short, the map of seismic gaps for Mexico and Central America done in 1973 needs to be updated periodically. Several of the gaps in figure 7.3 have not reruptured as of late 2018. With the knowledge that was available to us in 1973, it was not possible to pick which specific gap would rupture next. Nevertheless, the identification of several gaps narrows the choices of where to install instrumentation to "catch" a large earthquake and, it is hoped, to better understand their repeat times. Not knowing which gap will rupture next indicates to me that a good strategy is to install instrumentation in several gaps in the hope of recording a large earthquake and its precursors in at least one of them within a decade.

The U.S. government did not follow that strategy in the 1980s. Some of us proposed placing monitoring instruments in several segments of the San Andreas system in California that we regarded as more likely to rupture in large earthquakes in the following 20 years. Instead, the USGS placed all of our eggs in the Parkfield basket. Parkfield was a segment of the San Andreas Fault that had ruptured in several moderate-size earthquakes since 1857. The hope was to accurately forecast the fault's next shock within better than 10 years. When a predicted earthquake of magnitude 6.0 failed to occur on time at Parkfield, however, many scientists and public officials abandoned both earthquake prediction and intensive monitoring of several key areas in California and Alaska.

In 1978, I was invited to give the lead talk at a symposium in Guatemala City on the great, destructive Guatemalan earthquake of 1976. It was similar in magnitude and mechanism to the San Francisco earthquake of 1906 (and to the later shock on the Denali fault in central Alaska in 2002). We had an excellent bilingual field trip along the Motagua fault zone, which ruptured in 1976. The Motagua and one other major fault cut across Guatemala from east to west and form the boundary between the North American and Caribbean plates. Many Guatemalan students who previously had not visited the rupture zone, attended the symposium. I later flew to the ancient Mayan city of Tikal for a day.

South America

The Lamont group published several studies of seismic gaps for the very active and long plate boundary along the west coast of South America. Parts of the boundary are characterized by different maximum sizes of historic earthquakes. Many giant and great historic shocks have occurred there during the past 400 years, a number of which were very destructive. The long-term rate of plate motion along the entire subduction zone is high, about 2.6 inches (7 centimeters) per year. All of those factors make the South American subduction zone a good place to study long-term prediction. Figure 7.4 shows great earthquakes along that plate boundary from 1868 to 1978 and our mapping of major seismic gaps.

A great earthquake in 1906 of about magnitude 8.5 broke about 300 miles (500 kilometers) of the plate boundary off Colombia and northern Ecuador (figure 7.4). (This quake should not be confused with great earthquakes in California, Chile, and the Aleutians during 1906.) The southern part of that rupture zone broke again in 1942 (magnitude 7.7) and 1958 (magnitude 7.3). Those two shocks were not large enough to fill more than about 30 percent of the length of the 1906 rupture zone. Hence, the rest remained a seismic gap until an earthquake of magnitude 8.2 filled it in 1979. This is another example of *several* large earthquakes rebreaking a part of a plate boundary that had ruptured earlier in a *single* great event. The part of the 1906 rupture zone that broke in 1942 reruptured again in a shock of magnitude 7.8 in April 2016. Thus, it appears that the entire rupture zone of 1906 will not break in a single giant shock of magnitude near 8.5 but in a series of two or more great earthquakes.

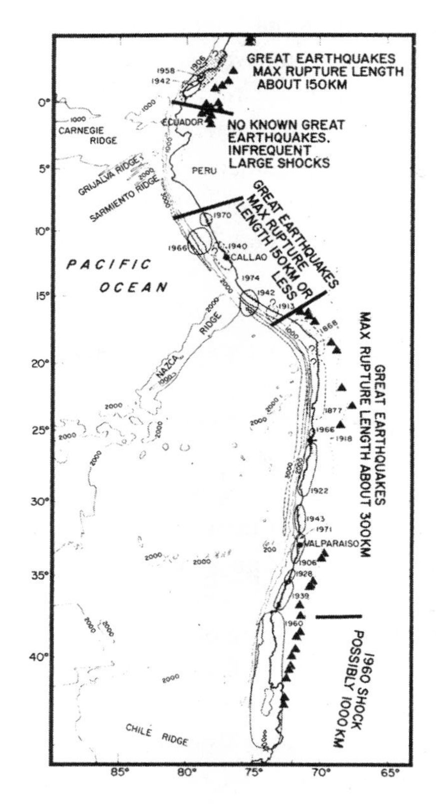

FIGURE 7.4 Rupture zones of large South American earthquakes between 1868 and 1978 (*shaded and open ovals*). Triangles indicate historically active volcanoes.

Source: McCann et al. 1979.

Great earthquakes have not occurred in the large seismic gap off southern Ecuador and northern Peru between about 1°S and 9°S (figure 7.4) during the existing historic record of about 400 to 500 years. The Carnegie and other ridges on the Nazca plate enter the subduction zone in that area. Because the rate of plate convergence in that region is high, either it is characterized by slow slip without great shocks, or the repeat time is very long. The slow-slip hypothesis, which some new GPS observations support, is probably correct. We do not understand how large topographic features on incoming plates with thick sediments, such as the Carnegie Ridge and the Tehuantepec Ridge off Mexico (near 95°W in figure 7.3), affect earthquake and subduction processes. Those gaps may never be the sites of great earthquakes. Additional geodetic data should be able to detect if those plate boundaries are in fact slipping slowly or building up stresses to great shocks.

I spent about a week in Lima, Peru, at an international seismological meeting in August 1973. It was winter there, and Lima was socked in with gray clouds. It is a huge city with a largely poor population. I participated in a weekend scientific excursion over the Andes via a narrow-gauge railroad from Lima. It took about fifteen hours to get to Huancayo, a city with a long-established seismograph station and magnetic observatory. En route from Lima, we climbed to higher than 15,000 feet (4.6 kilometers), where children were playing soccer. In one train car, most of us geophysicists remained in our seats near the top of the pass, but not Anton Hales of the Southwest Research Center in Dallas. One of the older scientists in our party, he walked around the car chain smoking.

Outside Huancayo, we were taken to a Spanish Catholic mission that had been the center for evangelizing the upper Amazon basin. Its church had a huge pipe organ, which one of my colleagues from Germany played. It was great to hear Bach in what was a very quiet setting. I then visited Equator for a week. Quito, the capital, is a beautiful Spanish colonial city surrounded by active volcanoes.

Two giant earthquakes ruptured the plate boundary along the coasts of southern Peru and northern Chile in 1868 and 1877 (figures 7.4 and 7.5). According to Chilean geophysicist Cinna Lomnitz in 1970, both shocks produced Pacific-wide tsunami. In 1979, Japanese seismologist K. Abe assigned a tsunami magnitude of 9.0 to each of those two events based on the sizes of their tsunamis. Abe calibrated the tsunami magnitude using twentieth-century earthquakes of known seismic magnitudes.

The 1868 and 1877 events very likely were giant earthquakes based on the sizes of their tsunamis and their long lengths of very strong shaking and damage parallel to the coast. Lomnitz states that the tsunami produced by the earthquake in 1868 carried several ships, including the U.S.S. *Wateree*, 3 miles (5 kilometers) inland, where it deposited them. The tsunami created by the giant shock in 1877 refloated the *Wateree* and moved it back closer to shore.

The deduced rupture zone of the 1868 earthquake, which is shown by a white line in figure 7.5 (and plate 6), was not the site of another great event until the shock of magnitude 8.4 off southern Peru in 2001 (purple line at upper left). The well-determined length of the rupture zone in 2001 (purple line) did not extend as far to the southeast as that inferred for the 1868 shock. Thus, the remaining seismic gap between about 18°S and 19.7°S still has a high potential for being the site of a future great earthquake. The 133 years

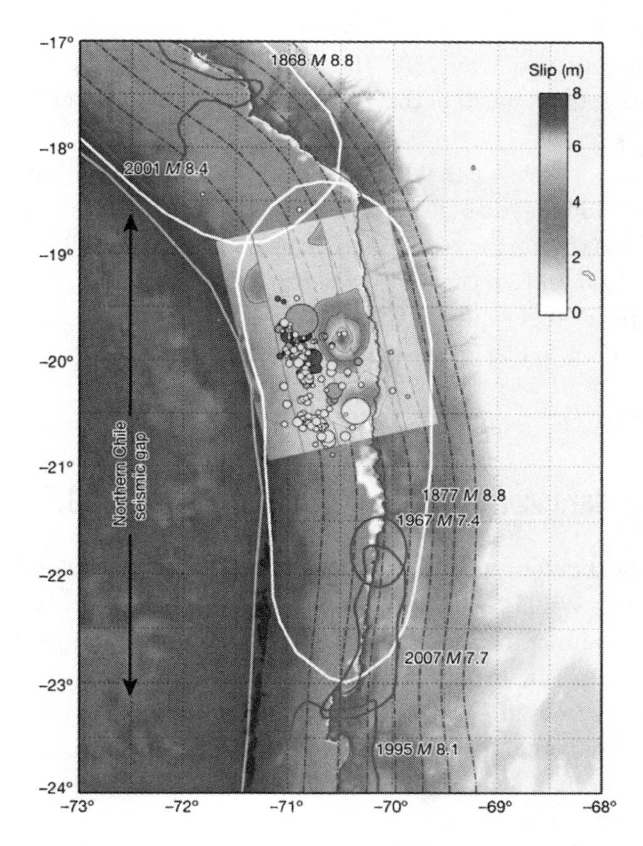

FIGURE 7.5 Contours of slip in the great earthquake of April 1, 2014, offshore of Iquique, northern Chile. Rupture zones of other large earthquakes since 1900 are outlined in the inset map. White lines enclose the inferred rupture zones of the giant shocks of 1868 and 1877. The large circle indicates the epicenter of the 2014 mainshock; darker circles denote its foreshocks; smaller light circles indicate events between the mainshock and the largest aftershock of magnitude 7.7 (*large light circle*); small light circles indicate subsequent aftershocks. See plate 6.

Source: Hayes et al. 2014.

between the shocks in 2001 and 1868 clearly indicates that zone was a long-standing seismic gap. The part that broke in 2001 no longer is part of the gap, of course.

The Tocopilla, Chile, earthquake of magnitude 7.7 in 2007 and the great Iquique shock of magnitude 8.1 in 2014 broke parts of the seismic gap still existing since the giant event of 1877 (figure 7.5). Segments of the plate boundary to the north and south of the rupture zone in 2014 continue to be seismic gaps as of later 2018. The moderate-size shock in 1967 was not large enough to fill much of the gap between the earthquakes in 2007 and 2014.

The parts of the zone that broke in 1877 subsequently had repeat shock times of 130 and 137 years (2007 and 2014), similar to the repeat time between the shocks to the northwest off southern Peru in 1868 and 2001. The earthquakes in 2007 and 2014 are discussed further in chapter 15. The part of the plate boundary that ruptured farther south in the great Antofagasta earthquake in 1995 also had been a long-standing seismic gap.

Our Incorrect Rankings of Seismic Gaps in 1979

I take a short detour here to examine estimates made in 1979 by Lamont seismologists William McCann, Stuart Nishenko, Janet Krause, and me of the likelihood that the rupture zones of the giant shocks of 1868 and 1877 in southern Peru and northern Chile would be the sites of large earthquakes in the few decades after 1979. We reexamined the seismic gaps described by Kelleher and others in 1973 and sought to identify rupture zones of earthquakes larger than magnitude 7.0 for many subduction zones.

The paper we published in 1979, "Seismic Gaps and Plate Tectonics: Seismic Potential for Major Earthquakes" (McCann et al. 1979), suffered from two problems. One was that the seismic and historic records of earthquakes of magnitude 7.0 to 7.75 and their aftershocks that we used to define their rupture zones were shorter than 50 years for most of the southwestern Pacific and the remote Cape Horn and South Sandwich regions. Hence, the sizes of seismic gaps and estimates of repeat times of large shocks for those areas were poorly defined. We were too ambitious.

We also made a mistake in estimating the chances that a number of seismic gaps that had ruptured previously in giant earthquakes would rerupture during the next few decades. We assigned our highest rating of seismic potential, category 1, to segments of plate boundaries that had not reruptured in great or giant earthquakes for more than 100 years—that is, not since 1879. We incorrectly took these segments to be the most likely sites of great shocks during the next few decades after 1979. Our category

2 included segments of plate boundaries that had experienced at least one large earthquake between 30 and 100 years earlier. We proposed that they had a significant but lower seismic potential than those in category 1. Those choices were wrong in 1979 and would be so today.

We soon became aware that shocks of magnitudes 7.0 to 8.0 occur most often along segments we assigned to category 2. None of the zones in category 1 that broke more than 100 years earlier was the site of a great or giant earthquake until the rupture zones of the giant earthquakes off southern Peru and northern Chile in 1868 and 1877 started to break again in 2001 and 2014. In addition, the sites of great-to-giant shocks along the Sumatran subduction zone of Indonesia in the late seventeenth and eighteenth centuries, which we assigned to category 1, remained quiet until the giant Sumatran shock of 2004. Since then, three other segments of the plate boundary off Sumatra have broken in great and giant earthquakes.

It was clear to us by 1982 that historic repeat times of many giant earthquakes were longer than 100 years because the average displacements in those events likely were unusually large. Plate motions could not restore the drops in stress during those giant events in less than about 100 to 150 years. We found that recurrence times of several shocks of magnitude 7.0 to 8.0 along very active plate boundaries were shorter. Displacements in them were smaller, and stresses dropped during them could be restored by plate motion within 30 to 100 years.

For example, displacements at the surface for the great southern California shock of 1857 averaged about 23 feet (7 meters). Dividing that displacement by the long-term rate of plate motion, 1.4 inches per year (3.5 centimeters per year, 3.5 meters per 100 years), indicated that about 200 years would be needed to bring stresses back to their pre-1857 levels. Paleoseismic work along those segments of the southern San Andreas Fault in the 1980s identified large prehistoric events with repeat times of roughly 200 years.

We also learned that the repeat times of great earthquakes varied considerably among and along various segments of active plate boundaries of the subduction and transform types. For some giant shocks, such as the Alaskan earthquake of 1964, paleoseismic studies identified repeat intervals as long as 700 years. Because giant shocks do not recur very often, the global occurrence of major to great earthquakes dominate a 30-year period. We learned something of value to long-term prediction.

Probabilistic Estimates of Repeat Times

Several of us soon realized that estimating the chance that major to giant shocks would recur during, say, the next 30 years along a given segment of a plate boundary would require information on either repeat times of past events along the boundary or how much stress remained to be built up since the boundary last ruptured in a large shock. It should be noted that the drop in stress at the time of an earthquake cannot be measured very accurately. Hence, we must deduce stress changes indirectly from displacements in large shocks divided by known rates of long-term plate motion.

In 1982, Stuart Nishenko, a graduate student at Lamont, and I began to make two types of calculations for segments of three very active faults in California. One method used average repeat times estimated from a combination of at least three historic and prehistoric large shocks at the same place. The other method calculated how much of the stress drop in the most recent large earthquake had been restored by slow plate motion. That method, which is called the *time-predictable model*, assumes that an earthquake recurs when the displacement in the most recent large shock is restored by plate motion. Our California forecast in 1984, which is described in the next chapter, included the likelihood or probability that a given fault segment would rerupture during the next 20 years. We calculated probabilities because the repeat times of past shocks and displacements along specific fault segments vary considerably along a major plate boundary.

I turn now to probabilistic calculations made by Nishenko in 1985 and to the identification of seismic gaps for the rest of the plate boundary along the west coast of South America.

Central Chile: Seismic Gaps from the Earthquakes of 1922 and 1943

Farther south along the coast of Chile between 26.5°S and 32°S (figures 7.4 and 7.6), most of the plate boundary ruptured in either the giant Atacama earthquake of 1922 (magnitude 8.5 to 8.6) or the great Illapel shock of 1943

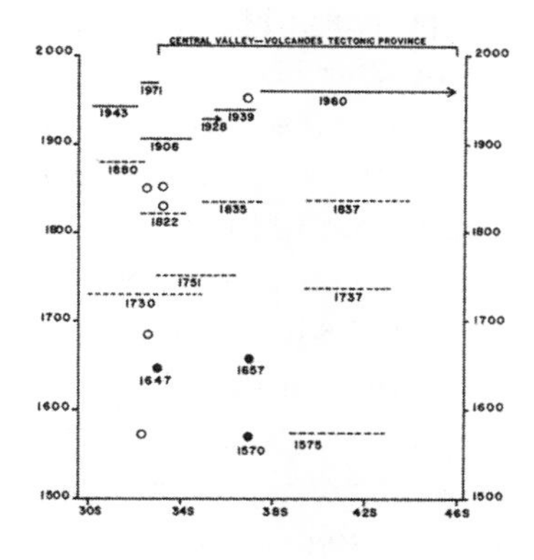

FIGURE 7.6 Rupture zones of great Chilean earthquakes from 1500 to 1970.

Source: Kelleher, Sykes, and Oliver 1973.

(magnitude 8.2). The gap from the shock in 1943 was filled by another similar earthquake of magnitude 8.2 in 2015.

A significant seismic gap remains from the giant shock in 1922. Nishenko states, however, that its rupture length was poorly determined. Lomnitz said it generated a tsunami that extended across the Pacific, indicating that it likely was a giant earthquake. Over the next few decades, much attention needs to be devoted to a giant earthquake that could rerupture the 1922 zone or to a series of great shocks that might break it.

Seismic Gap Following the Great Chilean Earthquake of 1906

The shock of magnitude 7.8 in 1971 broke only the northernmost part of the much longer seismic gap remaining from the great Valparaiso, Chile, shock of magnitude 8.0 to 8.5 in 1906 (figure 7.6). Nishenko reported that the rupture length of the Chilean shock between 32°S and 35°S in 1906 was well determined from observations of coastal uplift. He stated that great earthquakes in 1647, 1730, and 1822 ruptured approximately the same zone that ruptured in 1906. Nishenko combined those dates with uplift and inferred amounts of slip in 1906 to obtain an average recurrence time of 79 years. Using the dates of those three preceding great earthquakes, he predicted the recurrence of the shock to be 1910 ± 10 years. That 10-year uncertainty includes the date of the 1906 earthquake.

Nishenko stated in 1985 that the zone that ruptured in 1906 started to rupture again during the earthquake of magnitude 7.8 in 1971, and the rest of the zone was due for a great shock by 2005. He was interested particularly in a future great earthquake that would fill the remainder of the seismic gap of 1906. His forecast was successful in that in 1985 the great Valparaiso earthquake of magnitude 7.9 ruptured an additional part of that gap. The giant Maule earthquake of magnitude 8.8 in 2010 ruptured the remainder of the gap, but it also broke the adjacent seismic gap to the south. I return to the Maule event after discussing the giant shock that occurred to the south of it in 1960. The latter shock and its precursor one day earlier occurred in a zone that bears upon the earthquake of 2010.

Rupture Zone of Giant Chilean Earthquake of 1960

The shock of magnitude 9.6 in Chile in 1960 was the largest earthquake in the world to date since seismic records became available in the 1890s. It ruptured about 550 miles (900 kilometer) of the plate boundary between 38°S to 46°S (figures 7.6 and 7.7) and produced a Pacific-wide tsunami that was very large even in Japan. Previous earthquakes in 1837, 1737, and 1575 ruptured all or part of the 1960 zone. Lomnitz called them *Valdivia earthquakes*.

Kelleher, Oliver, and I concluded in 1973 and Nishenko concluded in 1985 that the shocks in 1837 and 1737 likely did not rupture the entire length of the 1960 zone. Based on paleoseismic data, Brian Atwater of USGS and his colleagues concluded in 2013 that the 1837 shock likely broke the southern half of the 1960 rupture zone. Data are especially poor for the 1737 event; it did not produce a damaging tsunami in either Hawaii or Japan. An earthquake in 1835 broke the northern 125-mile (200-kilometer) length of the 1960 zone (figure 7.7).

Lomnitz stated that the description of and extent of damage in the earthquake of 1575 were similar to those for the earthquake of 1960. Atwater and others found evidence that tsunami effects along the coast of southern Chile were similar for the two earthquakes. Nishenko obtained a repeat time of 211 years for the 1960 rupture zone by dividing its average displacement of 62 feet (19 meters) by the long-term rate of plate motion of 3.5 inches

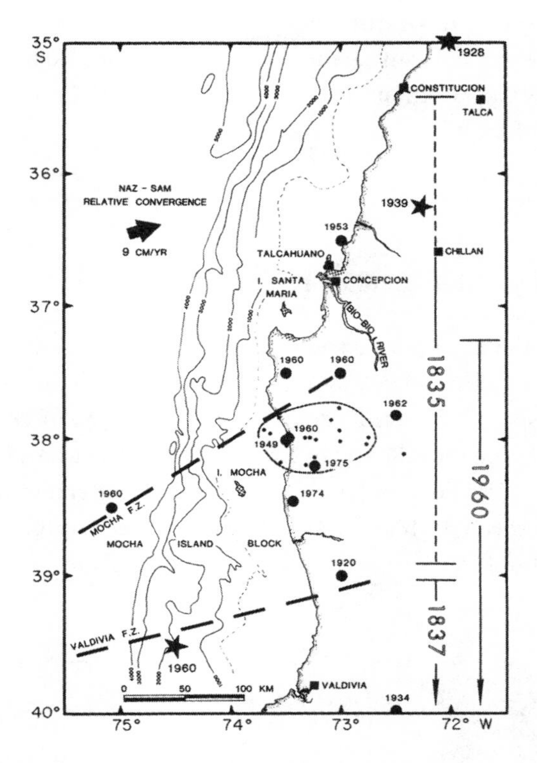

FIGURE 7.7 Rupture zones of great and giant Chilean earthquakes from 1835 to 1983 off southern Chile.

Source: Nishenko 1985.

(9 centimeters) per year. A more recent, better estimate of that rate, which is lower, gives a longer repeat time of about 270 years. The mainshock of 1960 also was accompanied by an inferred huge slow-slip event of magnitude 9.4. Since slip in the giant earthquake of 1960 was so large, it seems unlikely that it will recur during the twenty-first century. Smaller events like those of 1737 and 1837 may well rupture parts of the 1960 zone in less than 270 years.

The main earthquake of May 22, 1960, was preceded a day earlier by a foreshock of magnitude 7.9 in the Mocha Island block between 37°S and 38°S (figure 7.7). That block was the site of several shocks magnitude 7.0 to 8.0 between 1920 and 2011. The great shock of 1835 broke it and the plate boundary farther north. Likewise, the Mocha Island block ruptured a small amount in the giant Maule earthquake of 2010 (figure 7.8). Hence, the Mocha

Island block seems to rupture either along with adjacent great-to-giant shocks or independently in earthquakes of magnitude 7.0 to 8.0. It appears to be relatively weaker than the zones that broke with large displacements in 1960 and 2010.

Giant Maule Earthquake of 2010

Part or all of the plate boundary between about 35°S and 38°S (figure 7.8 and plate 7) ruptured in 1570, 1657, 1751, 1835, and again in 2010 during the giant Maule shock of magnitude 8.8. Although some aftershocks of the earthquake in 2010 (not shown) extended as far south as 38.5°S, calculated displacements (slip) greater than about 6.6 feet (2 meters) were confined to the region between 34°S and 38°S (figure 7.8). The rupture zone of 2010 has some, but not much, overlap with the rupture zones of the Valparaiso

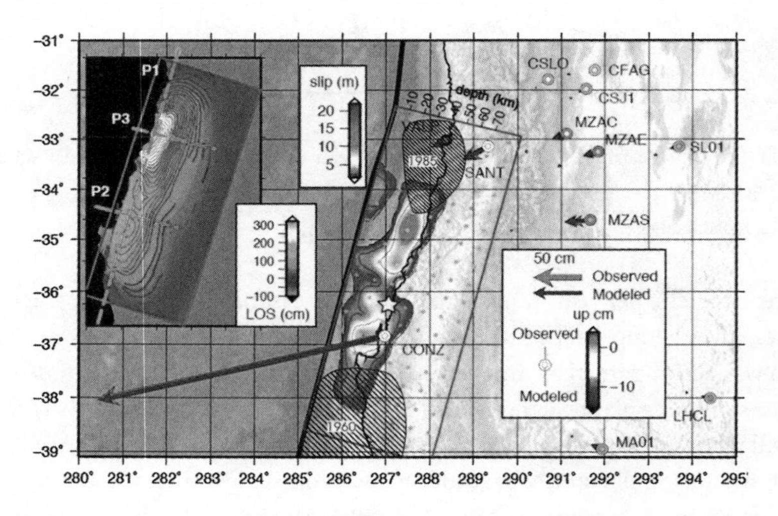

FIGURE 7.8 Slip during giant Maule earthquake of 2010 in central Chile, with contour lines from larger than 23 feet (7 meters) to about 3 feet (one meter). The large star denotes the epicenter, the point of initial rupture. The dots and arrows indicate amounts of slip at GPS stations. Hatched areas denote the rupture zones of the 1960 and 1985 earthquakes. The heavy black line is the axis of the Peru–Chile trench. See plate 7.

Source: Delouis, Nocquet, and Vallée 2010.

earthquake to the north of magnitude 7.9 in 1985 and the giant Chilean shock to the south on May 22, 1960.

The 1985 Valparaiso shock broke part of the seismic gap remaining from the great 1906 Chilean earthquake. The 2010 Maule shock ruptured the rest of the gap as well as the gap to the south that had previously broken in a great earthquake in 1835. Charles Darwin felt the shock at Valdivia in 1835 to the south of its rupture zone. He then visited the zone of strong shaking at Concepción, where he described uplift of the coast.

A tsunami generated by the event in 1835 flooded port facilities on the Chilean island of Juan Fernández 435 miles (700 kilometers) west of the coast of Chile in the Pacific Ocean. The earthquake of 2010 generated a tsunami that was destructive in Juan Fernández but not in Hawaii. The tsunami of 2010 was not as powerful as the tsunami generated by the Chilean earthquake of 1960. Only the northern part of the rupture in 2010 broke westward as far as the deepest part of the trench. Nishenko stated that the rupture length of the earthquake in 1835 (figure 7.7) was at least 120 miles (190 kilometers) and perhaps as long as 250 miles (400 kilometers). It is not clear how far north it ruptured. Whether it was just a great shock or a giant shock is not certain.

The shock of revised magnitude 7.7 in 1928, shown in figure 7.7, was not nearly as large as the earthquake of 1835, and its length was much shorter. The drop in stress in 1928 very likely was restored by slow plate motion well before the shock in 2010. The event of revised magnitude 7.6 in 1939 was located farther east and was a strike-slip shock that did not break the plate boundary. Hence, it did not fill part of the seismic gap left from the earthquake of 1835.

In retrospect, the plate boundary between 34.4°S and 37°S was a seismic gap for at least 50 years before the earthquake of 2010. Understandably, we did not realize this when we concluded erroneously in 1973 and 1979 (figure 7.4) that the shocks in 1928 and 1939 ruptured large parts of that plate boundary. We had used older seismic magnitudes of 8.3 for the shocks in 1928 and 1939. The revised and better-calibrated magnitudes that Javier Pacheco and I arrived at in 1992 and published in "Seismic Moment Catalog of Large, Shallow Earthquakes, 1900–1989" (Pacheco and Sykes 1992), indicate that those two events were, in fact, smaller.

In 1985, Nishenko stated that the plate boundary from the southern end of the rupture zone of the great 1906 earthquake near 35° S (figures 7.6 and 7.7) and as far south as the northern end of the earthquakes at 37°S in 1960 had a poorly constrained but possibly quite high potential for a series of large or great shocks between 1985 and 2005. Nishenko also used older larger-magnitude determinations for the 1928 and 1939 shocks, leading him to his more equivocal statement about the region's seismic potential between 35°S and 37°S. This illustrates the need to continue to work on the history and geologic evidence for historic and prehistoric earthquakes.

This is a good place to describe data from GPS satellites because they were important in understanding the Maule shock of 2010. During the past 30 years, GPS data have revolutionized the study of slip during and stress buildup to large earthquakes by providing precise information on relative movements near major plate boundaries—knowledge of which segments of subduction zones are fully locked and hence are building up strain prior to large earthquakes. GPS also provides detailed estimates of slip within the rupture zone of a single great shock. With repeat GPS observations, changes in horizontal distance between points hundreds to thousands of miles apart can be measured with an uncertainty of about 0.1 inches (a few millimeters). Dense arrays of GPS receivers are now in operation adjacent to several subduction zones and transform faults.

The 2010 Maule shock was the first large event for which GPS observations were made both during the preceding decade, during the earthquake itself, and afterward. Geophysicist Marcos Moreno and his colleagues of the Helmholtz Centre in Potsdam, Germany, along with scientists from France and Chile, deserve great credit for "staying the course" in making those observations, which they reported in 2010 (Moreno, Rosenau, and Oncken 2010).

Maximum slip in the Maule earthquake was about 66 feet (20 meters). The bold arrows in figure 7.8 show displacements at various GPS stations during that shock. The largest value was measured near the coast at Concepción (CONZ) just above one of the zones of inferred maximum slip. The GPS stations abbreviated SANT and VALP to the north of the main slip zones moved much smaller distances in 2010, indicating that little slip extended that far north. Those two stations were located above the zone that broke in the shock of 1985. The small displacements at GPS stations well east of the coast indicate that slip in the giant earthquake of 2010 was concentrated near and just off the coast along the plate boundary. Modeling of GPS data shows that slip in the mainshock of 2010 extended to a depth of 30 miles (50 kilometers).

GPS measurements before the Maule shock indicate that stations near the coast, such as CONZ, moved *eastward* in the decade prior to 2010 with respect to stations in the interior of the South American plate farther east. In contrast, CONZ moved *westward or seaward* during the earthquake. These movements indicate that the western edge of the plate near the coast was being *shortened as compressive stresses built up slowly prior* to the Maule shock. The region above the plate boundary *extended during the event in 2010 as compressive stresses were relaxed*. These observations are in good accord with the hypothesis that stresses are built up slowly and then released suddenly during great shocks.

8

THE SAN FRANCISCO EARTHQUAKE OF 1906 AND LONG-TERM PREDICTION FOR CALIFORNIA

My aunt Ethel Guildford lived in San Francisco for many years. A friend of hers who lived in the Bay Area during 1906 gave her a number of newspapers that were published soon after the great earthquake. I scanned them electronically and donated the originals to the New York Public Library. Some sensational headlines from two editions are reprinted in figures 8.1 and 8.2.

Spectacular fires burned a large area of what was then downtown San Francisco. Fires were spread first by the lack of water. Water mains to San Francisco from the Sierra Nevada were ruptured where they crossed the San Andreas Fault. When the U.S. Army lacked high explosives, it tried to stop the fires with gunpowder but instead spread them further.

EXTRA

Oakland Tribune.

PEOPLE SHOT DOWN BY SOLDIERS IN STREETS OF SAN FRANCISCO

SAN FRANCISCO, April 19.--It is estimated that at least 20 people have been shot here for stealing from unprotected stores and families and for insulting women. The city is under martial law and the soldiers do not hesitate, but shoot down anyone seen in the act of thieving.

PEOPLE BURNED ALIVE!

FIGURE 8.1 Headlines in the *Oakland Tribune* for April 19, 1906, the day after the earthquake.

EXTRA!

4.30 p. m. Oakland Tribune. 4.30 p. m.

STARVING DOGS
ARE DEVOURING
SCORES OF BODIES

CANINES DIG OPEN GRAVES

FIGURE 8.2 Headline in the *Oakland Tribune* for April 24, 1906.

The army shot many citizens either for looting or when they would not vacate their residences immediately. Many textbooks quote the number of deaths in the earthquake and fire as 600 to 700. Reanalyses indicate, however, much larger numbers—greater than 3,000 and perhaps greater than 5,000. For many decades, the State of California de-emphasized earthquakes and the number of fatalities in 1906. The headline in the *Oakland Tribune* about dogs eating bodies (figure 8.2) indicates that yellow journalism flourished in 1906.

In April 2006, my wife and I attended ceremonies and scientific talks in San Francisco for the one hundredth anniversary of that earthquake. Bells rang outside our hotel at 5:30 a.m. for the commemoration.

Likelihood of Earthquakes Occurring Along Major Faults in California

In 1982, Lamont graduate student Stuart Nishenko and I began work on long-term earthquake prediction for nineteen segments of the San Andreas, San Jacinto, and Imperial Faults, the three most active faults in California (figure 8.3). Figure 8.3 shows our results for the San Andreas alone.

We wanted to make quantitative steps in long-term earthquake prediction beyond merely identifying seismic gaps—that is, places where large shocks had not occurred for many decades or longer. We based our research on information from prehistoric large earthquakes that was just becoming available from geologic excavations of faults. We found that various segments of these three faults had ruptured in large earthquakes but of very different sizes or seismic magnitudes. Also, repeat times for large events along those segments varied considerably. The largest historic earthquakes along segments of the San Andreas Fault ranged from magnitude 6.0 to nearly magnitude 8.0. Because Nishenko and I could not use an average repeat time for, say, the entire San Andreas Fault, we divided the fault into a number of segments for analyses.

We realized that we needed more information to assess the probability that each segment would rupture in the future—within, say, the next 20 or 30 years. Hence, we set out to calculate the following for each segment:

1. The average recurrence time of past large shocks
2. How much time had elapsed since the most recent large event along each fault segment

Method 1 used dates of three or more past earthquakes to derive an average repeat time and its uncertainty. This technique is gradually improving as more information on prehistoric events is accumulating from paleoseismic (trenching) investigations.

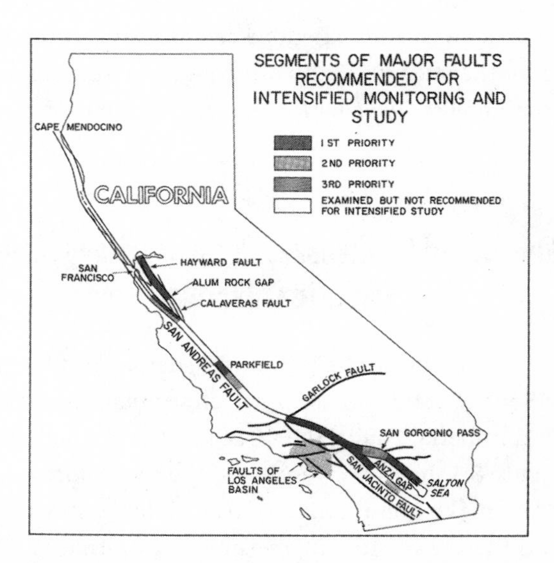

FIGURE 8.3 Major active faults in California and fault segments prioritized for special monitoring and study by the U.S. National Earthquake Prediction Evaluation Council in 1986.

Source: Unpublished Study by author, 1986.

For method 2, we calculated a repeat time of a future large shock along each fault segment from the average displacement (slip) in the most recent large earthquake divided by the long-term rate of fault slip. This method assumes that the next earthquake will occur when the reduction in stress (or strain energy) in the most recent large event is restored by slow plate motion and stress buildup.

We used the probability that an event will occur rather than claiming that large events either will or will not happen during the next 20 or 30 years. Weather forecasters regularly use probabilities to state that there is a certain percentage chance that rain will occur tomorrow. People do pay attention to high and low probability estimates: my house painters decided to stay home when the chance of rain the next day was 70 percent. Nishenko and I used the two methods to calculate the probabilities that each segment of those three very active faults in California would rupture in a large event from 1983 through 2002.

We defined large earthquakes (figure 8.4) as those that rupture the entire depth range from the surface to the maximum depth where slip occurs in earthquakes, typically 6 to 12 miles (10 to 20 kilometers) in California. Large earthquakes account for most of the slip along a fault segment and for most of the total energy release. Hence, we calculated probabilities only for large earthquakes. For the three faults in figure 8.3, almost all of the displacements are strike-slip and horizontal; the faults are nearly vertical.

Figure 8.5 shows calculated probabilities that various segments of the San Andreas Fault would rupture in a large earthquake between 1983 and 2002.

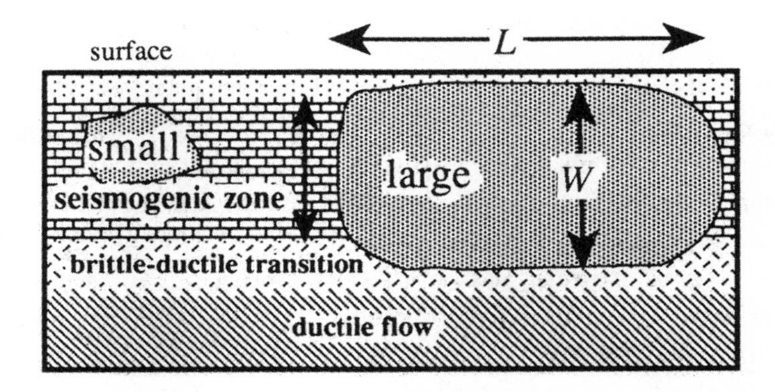

FIGURE 8.4 Two types of earthquakes along a plate boundary: large and small.

Source: Pacheco and Sykes 1992.

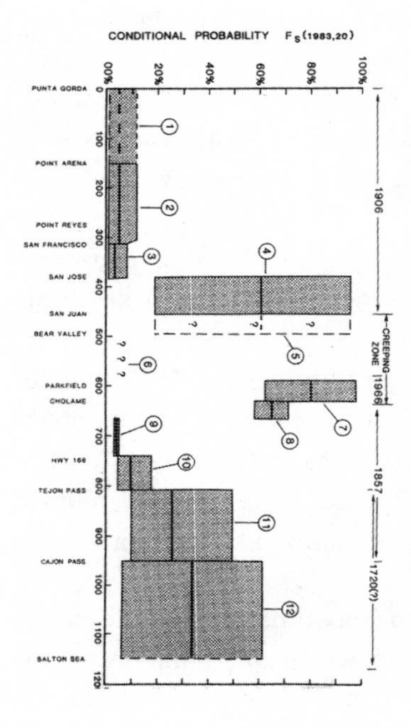

FIGURE 8.5 Probability on vertical axis that large future earthquakes would rupture segments 1 to 12 of the San Andreas Fault in California from 1983 through 2002. Horizontal axis extends at the left from the northwestern end of the fault at Punta Gorda to its southeastern end at the Salton Sea at the right.

Source: Sykes and Nishenko 1984.

Given the uncertainties in the estimates for each fault segment, it is most appropriate to express their calculated probabilities as very low, low, moderate, or high.

San Francisco and Northern California

The 1906 shock is often called the "San Francisco earthquake" even though it ruptured parts of the San Andreas Fault that were nearly 250 miles (400 km) in length from northwest to southeast of the city. Fault segments 1 through 4 (figure 8.5) broke together in an earthquake of magnitude 7.7. Because each of those four segments experienced different amounts of slip in 1906, Nishenko and I expected each of them to rupture next in large earthquakes at different dates. We corroborated this expectation using prehistoric dates obtained from paleoseismic investigations as well as historic dates. The loading of several fault segments by plate motion will occasionally be synchronized so that they rupture together, as in 1906. This synchronization also occurred in Colombia and northern Ecuador when three somewhat smaller but still large and great earthquakes subsequently broke parts of the rupture zone of the great Colombia–Ecuador shock of 1906 separately in 1942, 1958, and 1979 (figure 7.4).

Average slip along segment 2 to the northwest of San Francisco was about 13 to 16 feet (4 to 5 meters) in 1906, and its repeat time from paleoseismic observations was long, about 280 to 360 years. Thus, Nishenko and I calculated a low probability for its rupturing in a large shock from 1983 through 2002. We estimated that segment 4 south of San Francisco experienced only 2 to 4.6 feet (0.6 to 1.4 meters) of slip in 1906; we took its repeat time to be the 68 years between the earthquake in 1906 and the earthquake in 1838 that ruptured segment 4. Hence, we concluded that its likelihood of rupture in a large event was relatively high for the 20 years after 1983.

Figure 8.6 (plate 8) shows a view of northern California looking east across the San Andreas Fault where it passes on land near Point Reyes

FIGURE 8.6 View of the San Andreas Fault to the north of San Francisco (segment 2) looking eastward from the Pacific Ocean (*bottom*) to snow-capped Sierra Nevada peaks (*top*), n.d. Point Reyes Peninsula is at the bottom. The San Andreas Fault extends horizontally across the figure. See plate 8.

Source: Photograph courtesy of USGS, EROS Data Center, n.d.

Station and then underwater along Tomales Bay. Displacements in many past earthquakes and erosion of fractured materials within the fault zone have produced the long, narrow Tomales Bay. Recent probability calculations assign the zone a very low probability of rerupturing in the coming few decades—the same result Nishenko and I found for the period 1983–2002. Plate motion still has not restored stresses that were decreased in 1906 along segment 2.

Figure 8.7 shows me standing between parts of a fence that were offset in 1906 where it crossed the San Andreas Fault at Point Reyes Station. The well-known geologist G. K. Gilbert took the photograph in figure 8.8 in 1906 nearby, at the town of Olema. Slip in the 1906 shock created the fractured zone in the picture, which geologists call a "mole track."

One fascinating story that was repeated for decades involved a cow that supposedly fell into a mole track or fault opening at the time of the

FIGURE 8.7 The San Andreas Fault at Point Reyes Station where a fence was offset during the 1906 earthquake. The fence has been restored. See plate 9.

Source: Photograph of the author taken in 2006.

FIGURE 8.8 Rupture zone created by earthquake along San Andreas Fault in 1906.

Source: Photograph by G. K. Gilbert, plate 40 in Lawson [1908] 1969.

earthquake in 1906. News reports claimed that the opening then closed, leaving only the cow's tail visible at the surface. This story led many people to think that they could be swallowed by a fault during an earthquake. Many years later the brother of the farmer who owned the cow said the story was false. The cow had died the day before the earthquake, and the owner had simply pushed the cow into the newly opened mole track and covered it with dirt. When reporters sought exciting news, the farmer gave them the story about the cow being swallowed.

Loma Prieta Earthquake of 1989

Nishenko and I designated segment 4 of the San Andreas Fault as having a high but uncertain probability of rupturing in the 20 years after 1983 (figure 8.5). In October 1989, the fault broke in the Loma Prieta earthquake of magnitude 6.9. Rupture began with high-angle, dip-slip faulting, where the two plates moved toward one another, but it then propagated to the southeast as horizontal strike-slip motion along the vertical San Andreas Fault.

It is not surprising that the event in 1989 involved a combination of strike-slip and dip-slip faulting because the San Andreas Fault jumps to the west near the center of the Loma Prieta fault segment. Nearby Loma Prieta Mountain was formed by dip-slip faulting at what is called a compressional step-over or restraining bend along the fault. At that step-over, the San Andreas Fault is not a single plane but rather multiple planes of rupture. Step-overs develop with time to accommodate slowly changing plate motion with a minimum expenditure of energy.

In 1984, Nishenko and I correctly predicted the magnitude and time of occurrence of the earthquake in 1989 within our 20-year window, but we overestimated its length of rupture along the San Andreas Fault. In 1983, Allan Lindh of the USGS in Menlo Park, California, used similar probability calculations to predict the correct length and location, but he underestimated the magnitude as 6.5. At the time, he thought that part of the fault had broken previously in events of magnitude 6.5 in 1865 and 1890. Subsequent work by Martitia Tuttle of Lamont and me indicated those two smaller shocks likely did not occur on the San Andreas Fault but on other nearby faults.

In the 1980s, Wayne Thatcher and Michael Lisowski of the USGS used conventional geodetic data (pre-GPS) to predict that segment 4 was unlikely to rupture in a large event for several decades, at least for the remainder of the twentieth century. Christopher Scholz and Patrick Williams of Lamont separately argued, however, that segment 4 was a likely site of a large earthquake during the coming few decades. Scholz and Thatcher continued those earlier debates during sessions of the USGS Working Group on California Earthquake Probabilities in California in 1988, which are described in chapter 9. Contentious debate about the likelihood of large earthquakes centered more on segment 4 than on any other parts of the San Andreas Fault.

In the title of a lead article published in *Science* in 1990, the staff of the USGS at Menlo Park called the shock of 1989 an *anticipated earthquake* (USGS 1990a). In fact, although seismologist Lindh of USGS did anticipate the earthquake, Thatcher and Lisowski and some of their other colleagues in geodesy did not. A number of the Menlo Park geophysicists who failed to predict the 1989 event later implied that it had occurred off of the San Andreas Fault, which is not correct.

In 2007, Robert Twiss of UC Davis and Jeffrey Unruh of the consulting firm William Lettis and Associates published what is the most complete analysis of the 1989 earthquake using seismic, geologic, and geodetic data. For the northwestern half of the rupture zone, they show that the mainshock broke several strike-slip faults along the step-over zone of the San Andreas Fault. Their knowledge of structural geology was better than the knowledge of those involved in studying the earthquake at USGS.

Because an average of 6 feet (1.8 meters) of horizontal motion took place parallel to the San Andreas Fault in 1989, it is unlikely to be the site of a large shock for many decades. Geodetic measurements indicate that about the same amount of potential slip was built up as stress between 1906 and 1989. Hence, it is understandable that the fault was ready to rupture in 1989 and that this earthquake should not be regarded as a stray event located off of the San Andreas Fault.

Shaking in that earthquake collapsed a section of the San Francisco–Oakland Bay Bridge. The eastern section of the bridge was rebuilt recently in response to concerns about the old structure's earthquake safety. Most of the lives lost in this earthquake occurred when the Cypress Street Viaduct on the eastern side of San Francisco Bay collapsed (figure 8.9). The collapsed section was situated on weak bay mud. Damage within the city of San Francisco was high within the Marina District, an area of poor soil and debris dumped from the cleanup after the 1906 shock. Gas lines were ruptured, and fires broke out within that district. Buildings and homes there had also experienced strong shaking in the shocks of 1868 and 1906. In 1989,

FIGURE 8.9 Collapse of the Cypress Street Viaduct on the east side of San Francisco Bay during the earthquake in 1989. See plate 10.

Source: Photograph courtesy of Lloyd Cluff.

all of these structures were located about 60 miles (100 kilometers) from the shock.

Following the earthquake in 1989, the USGS formed a working group on the probabilities of future large earthquakes in the San Francisco Bay region. It reexamined estimated likelihoods for the San Andreas Fault for the following 30 years and calculated probabilities for other major faults in the Bay Area, including the Hayward, Calaveras, and Rodgers Creek Faults (figure 8.10). All of those faults are part of the San Andreas System of faults in the Bay area, of which the San Andreas Fault accounts for the greatest percentage of long-term slip between the North American and the Pacific plates.

The working group assigned a low probability (2 percent) of rupture to the northwest of San Francisco (segment 2), a moderate probability to the San Francisco Peninsula (segment 3), and a near zero probability to segment 4, the Loma Prieta rupture zone, which is also called the Southern Santa Cruz Mountains segment in figure 8.10.

I would now assign the San Francisco Peninsula segment 3 a moderate to high probability of rupture in an event of about magnitude 7.0 during the next 30 years. It ruptured in an earthquake of magnitude somewhat

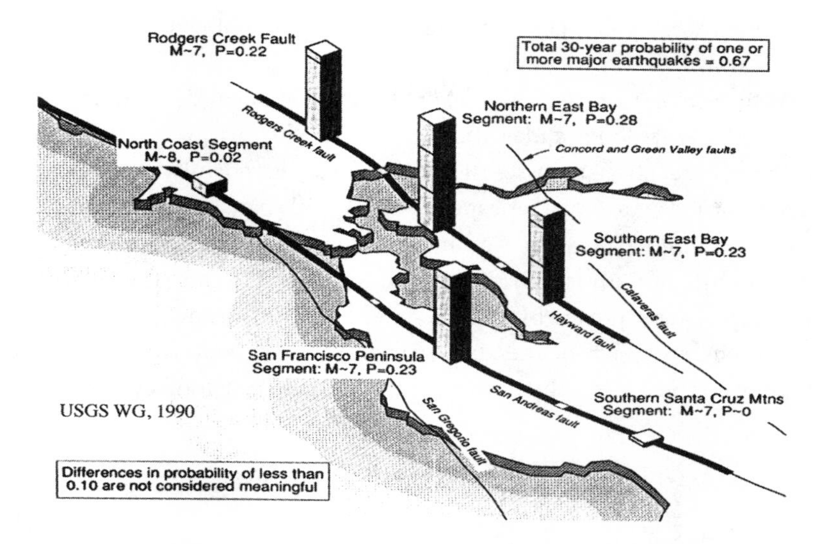

FIGURE 8.10 Probabilities of large earthquakes in the San Francisco Bay area from 1990 to 2020.

Source: USGS Working Group on California Earthquake Probabilities 1990.

larger than 7.0 in 1838 and again in 1906. Geodetic measurements indicate a lower probability. Compared to the Loma Prieta rupture zone, the San Francisco Peninsula segment of the San Andreas Fault, situated very near Silicon Valley and the city of San Jose, is located closer to many more people and assets.

Hayward Fault

Two segments of the Hayward Fault and most of the Rodgers Creek Fault have moderate probabilities of rupturing in large earthquakes between 1990 and 2020 (figure 8.10). The Hayward Fault last broke in a large event of magnitude 6.8 in 1868. It crosses through Oakland and several other densely populated cities on the eastern side of San Francisco Bay. Although the fault passes through the stadium at UC Berkeley, the rupture in 1868 did not extend that far northwest. A large shock on the Hayward Fault likely would be very damaging and costly, perhaps as great as $100–500 billion. Many people and structures are at risk.

Paleoseismic investigations by James Lienkaemper and his colleagues at the USGS in Menlo Park identified ten large prehistoric earthquakes prior to 1868 in a trench that crosses the southern Hayward Fault near Freemont. They obtained an average repeat time of 151 ± 30 years (Lienkaemper, Williams, and Guilderson 2010). Adding that repeat time to the date of the most recent earthquake gives a possible repeat in 2019 ± 30 years. The uncertainties in the dates reported by Lienkaemper and his colleagues for several of those prehistoric earthquakes, however, are large. In 2006, William Menke of Lamont and I used their data for the most recent four events, whose

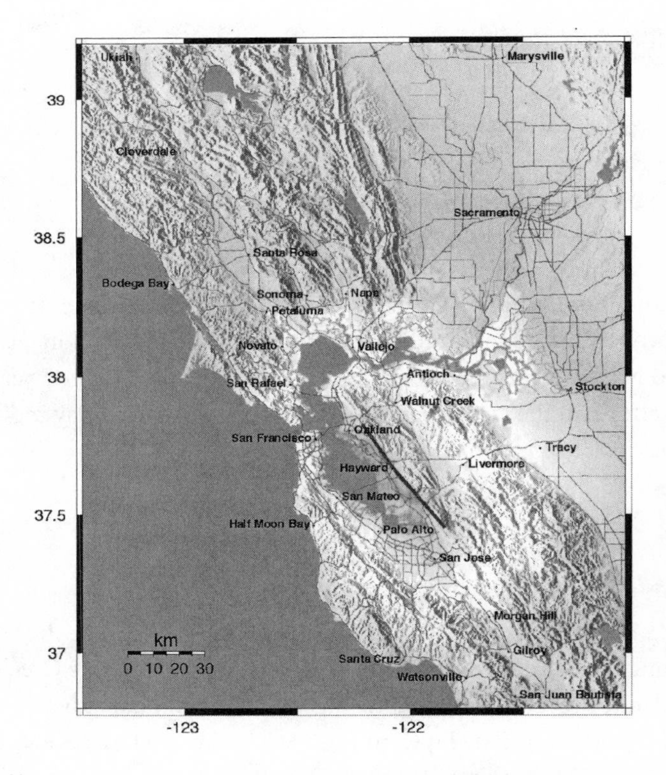

FIGURE 8.11 Projected levels of shaking in a future earthquake of magnitude 7.0 that would rupture the entire Hayward Fault (*heavy black line*). Darker shades indicate strongest shaking. See plate 11.

Source: USGS 2016.

dates are better determined, to obtain a shorter repeat time of 134 years (Sykes and Menke 2006). In either case, the Hayward Fault appears to be well advanced in its cycle of reloading to the next large earthquake because nearly 150 years have now elapsed since the last large shock in 1868.

Figure 8.11 (plate 11) illustrates calculations of shaking in a future earthquake of magnitude 7.0 that would break the entire Hayward Fault. Many areas of projected strong shaking are situated on bay mud and other poorly consolidated sediments along the shores of San Francisco Bay. Paleoseismic data on prehistoric earthquakes exist for only a single site on the Hayward Fault. The federal and state governments need to make a major, strenuous effort soon to obtain better information about past large earthquakes along the Hayward Fault and to be better prepared for a large and potentially destructive event.

Plate Movements, Slow Buildup of Stresses, and Displacements During a Large Earthquake

This is a good point to describe displacements at the surface—that is, slip—during a large earthquake, slow stress buildup beforehand, and long-term plate motion as averaged over many large events. Figure 8.12 illustrates these concepts using three hypothetical map views of the San Andreas Fault.

In 1908, Harry Fielding Reid of Johns Hopkins University formulated his now famous theory of slow stress buildup beforehand and then sudden stress drop at the time of the earthquake of 1906. Reid used geodetic surveys of horizontal positions (latitude and longitude) on the surface of the earth made at two times in the 50 years prior to this shock and once soon afterward.

Geodetic surveys of markers driven into the ground were made to assist in the subdivision of land soon after California became a state. The markers, which were placed about 20 miles (30 kilometers) apart, moved with respect to one another during the 1906 shock. Because surveying was an expensive job, the state and the U.S. governments were annoyed that they needed to resurvey those geodetic markers so boundaries of property and of towns and counties could be determined with confidence. Although not conducted for scientific understanding, these resurveys were the most

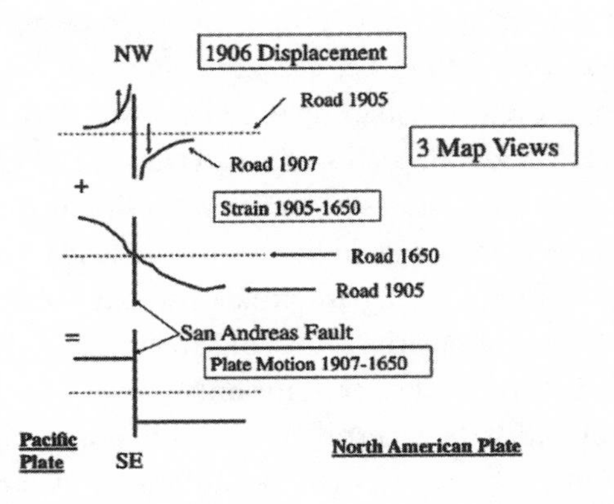

FIGURE 8.12 Displacements of a hypothetical road that crossed the rupture zone of the earthquake in the San Francisco Bay Area in 1906 at three times intervals. Each of the three panels is arranged with the vertical axis oriented parallel to the fault (*solid black line*)—that is, northwesterly. Each horizontal axis extends northeast to southwest perpendicular to the San Andreas Fault. The top panel shows the displacement of the earth during the 1906 shock (1907–1905). The middle panel shows how the road was bent (and the region was strained) between just after the last great shock about 1650 and just before the 1906 earthquake in 1905. The bottom panel shows that the combined displacement in the top two panels is equal to the plate motion from just after the 1650 event to just after the 1906 shock.

Source: Unpublished figure by the author, 2010.

accurate geodetic surveys at the time. Today, this job can be performed much more easily and inexpensively with GPS.

A road in the top part of figure 8.12 illustrates the displacements Reid found in his analysis of the geodetic data from before to just after the earthquake in 1906. The road shown by a dotted horizontal line in 1905 was offset in 1906 to the positions in 1907 indicated by the two heavy solid lines. Reid found that the offset during the 1906 earthquake was a maximum at the San Andreas Fault (vertical black line) and decreased to zero at distances of about 25 miles (40 kilometers) away from it.

Although Reid possessed data going back only to 1857, I have extrapolated his findings to just after the previous large 1906-type earthquake, which paleoseismic work indicates occurred around 1650. The middle panel in figure 8.12 shows how a road that was straight just after the earthquake in 1650 would have been bent—that is, strained or distorted—by 1905 just before the shock in 1906. Note the road was not offset at the fault during that interval and that the San Andreas Fault itself was "locked." Fault roughness prevented slip from occurring at the fault during this long period of stress and strain buildup. The combined displacements in 1906 plus those between 1650 and 1905 equal the total plate movement in the lower panel of the figure.

Southern California

In 1857, a great earthquake of about magnitude 7.8 ruptured segments 7 through 11 of the southern San Andreas Fault (figure 8.5). Because this earthquake occurred well before seismic recordings, Nishenko and I estimated its magnitude from the surface displacements mapped along those fault segments by Kerry Sieh of Cal Tech and the depth of rupture using the maximum depths of well-located recent small earthquakes.

We computed a moderate probability that segments 11 and 12, which experienced large amounts of slip at the surface in 1857, would rupture again between 1983 and 2002. If they broke separately, either could be the site of an earthquake of about magnitude 7.5. Neither has ruptured as of late 2018.

In 1978, Kerry Sieh published his classic study of prehistoric earthquakes at Pallett Creek (PC in figure 8.14 and plate 12), which is located along segment 11 of the San Andreas Fault to the northeast of Los Angeles. Sieh identified about seven prehistoric earthquakes at Pallett Creek. The most recent shock cuts all of the sediments, whereas the older earthquakes cut only older sediments that existed at the time (figure 8.13 and plate 11).

Sieh obtained the approximate date of each prehistoric event using the radioactive isotope carbon-14. He dated peats in the sediments just above and just below each earthquake that displaced the walls of his trenches. Peat consists of carbon from decayed wood and animal remains. Carbon-14 is generated in the upper atmosphere by the collision of cosmic rays with atoms of carbon. If it is incorporated quickly into streams and buried as peat, as at Pallett Creek, its date is fairly accurate.

Paleoseismic dates, however, are not very accurate for some other sites in the deserts of southern California, where the carbon-14 deposited may have been created by a fire, say, 50 to 100 years before an earthquake and

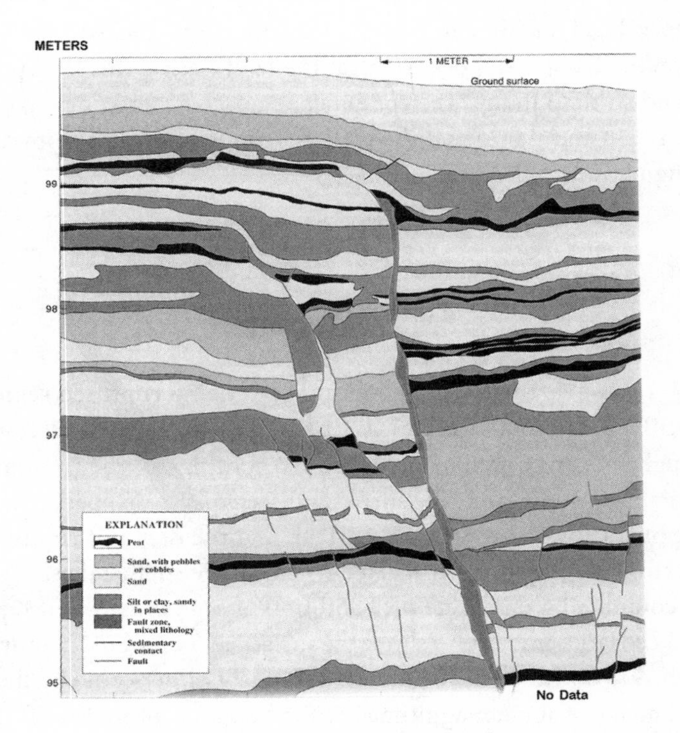

FIGURE 8.13 The San Andreas Fault exposed in the southeast-facing wall of a trench excavated by Kerry Sieh in the 1970s at Pallett Creek. The fault zone is indicated by the vertical shaded area; individual fault breaks in three past earthquakes are indicated by vertical lines. The excavation extends from the surface to a depth of about 16 feet (5 meters). The Pacific plate is to the left, and the North American plate is to the right. See plate 12.

Source: USGS 1990b.

not deposited as peat. This likely affected the dates of paleo-earthquakes at Wallace Creek (WC in figure 8.14).

Figure 8.14 indicates the rupture zone of the great earthquake of 1857 as mapped by Sieh as well as the zones deduced for two previous great shocks along the San Andreas Fault in southern California. Sieh and Gordon Jacoby of Lamont set out to ascertain if the shock in 1857 disturbed the growth of trees along the fault near Pallett Creek. Jacoby, who died in 2014, was one of the two founders of the tree-ring laboratory at Lamont. Although Sieh and Jacoby did find a disturbance in the tree rings for 1857, they found a larger disturbance for 1812. Provided that the rings of trees of a given

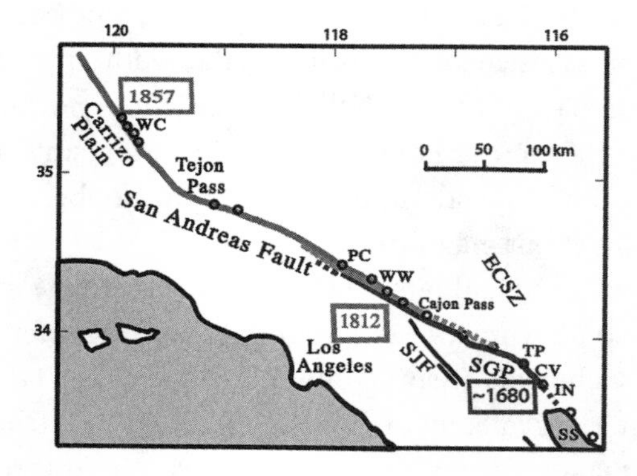

FIGURE 8.14 Parts of the San Andreas Fault that ruptured in the great earthquakes of 1857, 1812, and around 1680 in southern California. PC is the site of the trench at Pallett Creek shown in figure 8.13. Open black circles indicate paleoseismic (trenching) sites. See plate 13.

Source: Sykes and Menke 2006.

species are widely sampled in a region, as in much of southern California, dates can be determined to within one year.

A great earthquake in 1812 was widely felt in southern California. The tree-ring analysis showed that it occurred on the southern San Andreas Fault, a determination that was not clear prior to that work. Subsequent paleoseismic investigations indicated that the 1812 shock ruptured at least the portions of the fault labeled green in plate 13. The displacement in the 1857 earthquake was only 3 to 6 feet (1 to 2 meters) at its southeastern end near Pallett Creek and Wrightwood (WW in figure 8.14). The displacement in trenches at Wrightwood in 1812 was much larger than the displacement in 1857, as was the displacement in a previous great prehistoric earthquake that occurred around 1680 (blue in plate 13).

Paleoseismic investigations indicate that great earthquakes occur about twice as often at Pallett Creek and Wrightwood than at the northwestern end of the 1857 rupture zone. This makes plate tectonic sense in that displacements in past great earthquakes at Pallett Creek and Wrightwood are about half the displacements to their northwest in the Carrizo Plain. Thus, the long-term rate of plate motion is about the same at those sites.

The southernmost segment of the San Andreas Fault in California has not broken in a great earthquake since around 1680. This nonactivity has puzzled many earth scientists, including me. More recent GPS and paleoseismic work indicates that the San Jacinto Fault (SJF in figure 8.14) experiences about as much long-term slip as the southernmost San Andreas Fault. Hence, that part of the San Andreas Fault does not rupture as often as segment 11 to the west at Pallett Creek. The San Jacinto Fault ruptures more often than the San Andreas but is not known to break historically in events larger than about magnitude 7.4.

In addition, a significant but controversial percentage of the displacement along the southernmost San Andreas Fault is transferred northward into the eastern California seismic zone (ECSZ in figure 8.14), bypassing the complex tectonic knot in San Gorgonio Pass (SGP). Nevertheless, although the great earthquake of around 1680 probably ruptured through San Gorgonio Pass, not all shocks seen in the trenches at Pallett Creek and Wrightwood have done so. Present scientific research focuses on determining the modes and dates of faulting in great earthquakes that break through San Gorgonio Pass.

Parkfield Earthquakes

In the 1970s and 1980s, segment 7 of the San Andreas Fault at Parkfield (figure 8.5) was thought to have broken in six earthquakes of only about magnitude 6.0 between 1857 and 1966. The Parkfield segment is located between the northwestern end of the rupture zone of 1857 and segment 6, which has no history of large earthquakes and is creeping (slipping slowly with time) at the full rate of plate motion. In 1982, Nishenko and I computed very high probabilities that the Parkfield segment would rerupture between 1983 and the end of 2002. It did not rupture by 2002, but it broke in a shock of magnitude 6.0 in 2004. A more specific prediction for Parkfield by USGS scientists is discussed in the next chapter.

None of the other segments of the San Andreas, Imperial, and San Jacinto Faults, which we calculated as having a low probability of rupture, broke during our forecast time window from 1983 to 2002. If any one of them had broken, it would have falsified our methodology of long-term prediction. We cannot estimate which of the segments in northern and southern California that we assigned a moderate probability will be the next to rupture. More recent working-group reports on California include additional faults. If any one of these faults breaks, especially in an urban area, it is likely to cause considerable damage.

Southern California Earthquake Center

In 1987, the U.S. National Science Foundation held its first competition for major funding of Science and Technology Centers. Cal Tech and the University of Southern California (USC) submitted separate large proposals for work on earthquakes, but neither was funded. In the summer of 1988, I attended a U.S.–Japan meeting on earthquake prediction at Morro Bay, California, where I suggested to Robert Wesson, the head of seismology, volcanoes, and engineering at USGS, that a great need existed for a consortium made up of universities, USGS, and the Geological Survey of California to form a "center without walls" for studies and monitoring of earthquakes in southern California. I told him that the director of the proposed center needed to be from a university in southern California and that Keiiti Aki of USC would make an excellent choice.

Wesson agreed with my suggestions for a center and for Aki as its director. We talked with Aki at the meeting, and he strongly favored my proposal. I soon left the United States for a planned five-month sabbatical in New Zealand. Aki held a workshop to create a center, then submitted a proposal for it to the National Science Foundation's program for Science and Technology Centers. His group came up with a better name than I had—the Southern California Earthquake Center (SCEC) instead of Center for the Study of Earthquakes in Southern California. It received governmental funding in February 1991 and continues today.

When SCEC was being formulated, major issues included which universities would be involved and funded. Several scientists from three southern California universities wanted all the funding to go to them, believing they had all of the expertise that was needed. Aki was more cosmopolitan in his views and wanted to include the best scientists regardless of the locations of their institutions. He also thought that SCEC should be a national demonstration project that might be duplicated elsewhere. Aki prevailed, and about seven universities, including Columbia, became core participants in SCEC. I was involved until just before retiring in 2005.

We decided that tenured professors like me would not receive salaries from SCEC so that more funds could go to the support of graduate students and young researchers. Two of my Ph.D. students, who were funded by SCEC, worked on earthquakes and the evolution of stresses in southern California. Several members of the research staff at Lamont were also involved and supported. The formation of SCEC allowed universities to compete effectively with USGS for work on and research funds for the study of earthquakes in southern California. Following Aki's departure from SCEC and USC in 1995, a more provincial view ensued at SCEC, and Lamont received fewer funds. Aki died in 2005.

Alaska Volcano Center

A few years after 1988, I attended a meeting in Alaska organized by John Davies, the state seismologist, about proposed plans for studying earthquakes and volcanoes in Alaska. At the meeting, I suggested forming the Alaska Volcano Center to study and monitor active volcanoes to the west of Anchorage, the most populous city in Alaska. Ted Stevens, a powerful senator for Alaska at the time, arranged for funding to come through USGS with the stipulation that at least half would go to the State of Alaska. The center has done well and is continuing.

An unexpected benefit of the Alaska Volcano Center is its issuance of warnings of volcanic eruptions in Alaska and the Aleutians to aircraft flying to and from the western Pacific. Air traffic on those routes has increased dramatically in the past several decades. During major eruptions, those volcanoes typically eject huge amounts of dust and other debris that can disable the engines of jet aircraft that inadvertently fly through them.

9

MY WORK WITH THE U.S. NATIONAL EARTHQUAKE PREDICTION EVALUATION COUNCIL

The United States Geological Survey is charged with evaluating and making official predictions of earthquakes in the United States. In 1979, it set up the National Earthquake Prediction Evaluation Council (NEPEC), to advise the director of the USGS about the scientific validity of earthquake predictions.

NEPEC consists of ten to fourteen members, about half of whom are USGS employees and the other half from universities, state governments, and consulting firms. I was a member of NEPEC from 1979 to 1982 and served as chair from the fall of 1984 until I went abroad in mid-1988. During my tenure as chair, the state geologists of California and Alaska were members. NEPEC discussed only U.S. earthquakes, except when we were requested to review the Brady-Spence prediction of a supergiant earthquake along the subduction zone of Peru and Chile, which did not occur.

In 1984, John Filson, chief of the USGS Office of Earthquakes, Volcanoes, and Engineering, asked me to chair NEPEC after reading my paper with Stuart Nishenko on the probabilities of earthquakes along three active faults in California (Sykes and Nishenko 1984). Prior to 1984, NEPEC was charged only with evaluating predictions that were made by various parties. I thought, however, that a national in-house forum was needed to evaluate the likelihood that some regions of the United States could be the sites of large and damaging earthquakes on time scales of a few decades. In my letter of appointment in August 1984, Dallas Peck, director of USGS, agreed with my recommendation that "NEPEC should take a broader view in advising the Director on earthquake hazards. It should take an active role in assessing earthquake potential in various regions rather than just the passive role of responding to predictions by other parties. NEPEC should meet two to four times each year."

While I chaired NEPEC, we reviewed the San Francisco Bay area, Parkfield, southern California, southern Alaska, and the Cascadia subduction zone of Washington, Oregon, and northernmost California. We examined some areas in California more than once. For each of our discussions, which typically lasted about two days, we invited knowledgeable scientists to speak to NEPEC on results relevant to the area we planned to discuss.

I asked speakers to arrive with an abstract and copies of figures for their presentations. These documents were included in a summary of the meeting sent to the USGS director and in a quick publication as a USGS Open File Report. I followed the practice of quick publication established by the Japanese Coordinating Committee for Earthquake Prediction. It provided information about new developments to individuals in the public, state, and federal governments and to others in the scientific community. Although not ready to make specific predictions, many scientists did not want to sit on data and results that might be relevant to possible future damaging earthquakes. They also were less inclined to leak their results to the press when they had a chance to speak to NEPEC and have their results published quickly.

Announcements to the Press: Radon Measurements in Southern California

When I agreed to chair NEPEC, I was concerned about several previous evaluations of earthquake predictions in the United States as well as about the role of public-relations activities by USGS staff. In November 1981, a data review meeting had been held at Cal Tech to discuss what appeared to be anomalous amounts of the gas radon in a 60-mile-long (100-kilometer-long) region of southern California and its relationship to either a possible future earthquake or other tectonic deformation. The USGS public-relations office issued a press release about the meeting. About one hundred members of the media showed up anticipating that a major announcement would be made about a coming large earthquake. A written statement by the group at the Kellogg Radiation Lab of Cal Tech, which observed the radon, was fairly circumspect, saying, among other things, "The possibility that they [the radon anomalies] are premonitory must be considered," and calling for additional geodetic measurements and prompt analysis of seismic data.

Members of the media were disappointed when various scientists spoke at the meeting about what the media considered to be fairly esoteric topics in seismology, geodesy, and radon. The press then cornered individual scientists in the hallways, essentially asking them, "Where's the beef?" and "Is something being withheld?" It was not a good forum for either members of the media or scientists.

Brady-Spence Prediction of a Supergiant Earthquake

Another public-relations disaster was a USGS press release in 1981 stating that NEPEC would meet that January to evaluate a prediction by Brian Brady of the U.S. Bureau of Mines and William Spence of USGS regarding either a catastrophic supergiant earthquake or a series of giant events that would rupture the seismic gaps that most recently broke during the giant earthquakes of 1868 and 1877 in southern Peru and northern Chile (figure 7.5). Brady had traveled to Peru to inform its president that he was sorry, but a giant earthquake would occur within a few days of either August 10 or September 16, 1981. The president requested that the USGS director evaluate the prediction, which he forwarded to NEPEC.

I was a member but not the chair of NEPEC at the time. Walter Cronkite, the anchor of *CBS News*, sent a team to film the proceedings. His party and others from the media assumed that a prediction of a great earthquake would be confirmed by NEPEC. This did not happen. Spence and those of us on NEPEC were aware that the recurrence of two giant earthquakes more than 100 years earlier along that part of the subduction zone was the basis of Brady's short-term prediction.

Brady's scientific work was well known to several of us on NEPEC. Some of us were also aware of his attempts to predict mine bursts—that is, small earthquakes in deep mines. Nevertheless, we were highly skeptical of his attempt to predict one or more giant earthquakes in South America within several days more than seven months ahead of time. Although Spence wanted to make sure that Brady's work was heard, in June 1981 he withdrew his support for Brady's prediction after attending the NEPEC review. A huge amount of money was spent on travel, the NEPEC meeting, and the salaries of USGS and other federal employees who attended the meeting. That money could have been better spent on support of students, monitoring, and research.

Unfortunately, the chair of the panel, Clarence Allen of Cal Tech, negotiated with Brady that he would have one and a half days to make his presentation. Allen informed the rest of us on the panel about his agreement with Brady when we arrived in Colorado for the review. Ten minutes into Brady's talk, it was clear to me that his methodology was flawed and his prediction was not valid. The panel's report reflected my views, which was conveyed to the USGS director and the government of Peru.

The CBS crew went home empty-handed. They managed to film several of us yawning on the first morning of Brady's presentation; fortunately, we were not shown on the evening news. Filson of USGS agreed to travel to Peru and to be present at the times of Brady's prediction. Neither a great earthquake nor a giant earthquake occurred during those times. Three great, but not giant, earthquakes occurred in those two seismic gaps in 2001, 2007, and 2014, as described in chapter 7. Other parts of those gaps have not yet broken again.

Paul Krumpe of the Office of Foreign Disaster Assistance of the U.S. Agency for International Development (USAID) told some members of the press during the NEPEC meeting in 1981 that Brady's work was on a par with that of Isaac Newton. Krumpe was responsible for distributing funds from USAID for earthquake risks in Latin America. His views reflected badly on USAID's scientific competence.

Over time in my involvement with NEPEC, I also found that the U.S. Federal Emergency Management Agency (FEMA) lacked both interest and scientific competence in earthquakes. No one from FEMA attended NEPEC meetings even when they were specifically invited and even though FEMA was responsible for coordinating the National Earthquake Hazards Reduction Program. Congress later removed FEMA as head of that effort.

Quick Publications and Dealing with the Media

I learned valuable lessons from the Brady prediction and the discussion of the radon anomalies in southern California. Brady had neither a manuscript nor even an abstract of the prediction he presented to us in Colorado. As chair of NEPEC, I insisted that speakers at our meetings arrive with abstracts and figures of either their predictions or presentations. Several authors had changed their predictions over time, so I felt it was important to have a written record of what was actually presented and evaluated by NEPEC.

When predictions of earthquakes as small as magnitude 4.5 were brought to NEPEC, we decided that although U.S. earthquakes smaller than magnitude 5.0 could be of scientific interest, they were unlikely to cause damage or injury. Thus, they should neither be announced to the public nor conveyed widely to governmental officials. Nevertheless, they remained in our published minutes.

Many individuals sent predictions to either USGS or me as chair of NEPEC. The council agreed that I would distribute copies to all members of NEPEC. If a single member wanted to have a prediction discussed, we would do so at our next meeting. We thus dispensed with several "quack" predictions when no members opted to discuss them.

The NEPEC meetings were listed beforehand in the *Federal Register*. Filson agreed with my suggestion that USGS should not issue a press release for each meeting. We kept the presentations for the first day and a half open to the public, including the media, as required by law. No members of the media attended any of approximately ten meetings I chaired. We held a closed executive session at the end of each meeting to discuss our evaluation of any predictions we heard as well as of presentations of areas under review. I think these policies allowed NEPEC to conduct its work effectively, with scientific credibility, and to use its members' time wisely. Our procedures allowed scientists to come forth with their results to a group of their peers without undue publicity and public confusion.

In 1986, after two years of work with me as chair, NEPEC recommended for special study several faults in California, including some the Los Angeles area (figure 8.3). The council members were unanimous in their assessments of all fault segments except for what became the Loma Prieta rupture zone of 1989 (segment 4 of Sykes and Nishenko 1984). Two areas in Alaska that we examined were recommended for special study but not at the highest priority.

In 1986 and 1987, Christopher Scholz and Wayne Thatcher continued their previous debate about the likelihood of a large earthquake along segment 4 of the San Andreas Fault. The slip in 1906 was larger to the northwest of San Francisco than the slip along the 85-mile (140-kilometer) length of segments 3 and 4 (figure 8.5) of the fault to the southeast. Hence, stresses that had decreased in 1906 along the San Francisco Peninsula, segment 3, and along segment 4 were more likely to be restored sooner. In addition, Thatcher and Scholz agreed that the amount of slip in 1906 decreased substantially between Mussel Rock in western San Francisco at the northwestern end of segment 3 and the southeastern end of segment 4 at San Juan Bautista. Hence, a future large earthquake was most likely to occur sooner along segment 4 than along segment 3.

Thatcher and Scholz, however, disagreed about the probability that segment 4, which was 45 miles (75 kilometers) long, would rupture sometime during the remainder of the twentieth century in an earthquake of about magnitude 7.0. Thatcher argued several times that a geodetic estimate of slip of 8.5 ± 1 feet (2.6 ± 0.3 meters) along segment 4 in 1906 was more reliable than the smaller measurements of displacement at the surface. Scholz stated in several letters that the geodetic estimates were critically dependent on the measurements to and from a single geodetic marker on Loma Prieta Mountain near the fault. He claimed that the geodetic network for segment 4 was sparse and that several measurements were more uncertain than Thatcher thought. Scholz also emphasized the smaller fault offset of 5 feet (1.5 meters) made in a tunnel at depth, which extended through the fault, and concluded that it was a more accurate, albeit sparse, estimate of slip in 1906.

Scholz's estimate of horizontal slip in 1906 for the Loma Prieta rupture segment was nearly identical to the slip measured for the 1989 earthquake. The 83 years between the two earthquakes was a sufficient time to build up enough stress to generate the average strike-slip displacement of about 5 feet (1.5 meters) that occurred in 1989. Scholz and Thatcher were unable to obtain geodetic measurements of stress buildup between 1907 and the 1980s and to continue their analyses and debate prior to the Loma Prieta shock of 1989.

The 1989 event broke 28 miles (45 kilometers) of the 47-mile (75-kilometer) length of segment 4. The moderate-size Chittenden earthquake of 1990 ruptured an additional 7.5 to 10 miles (12 to 15 kilometers) of the adjacent segment to the southeast. Data for the 1989 shock, of course, were much better than those for the 1906 earthquake. About 10 miles (15 kilometers) of the northwestern end of segment 4 did not rupture in either 1989 or 1990. That remainder is located at the northwestern end of the complex step-over of the San Andreas Fault, where it is difficult to determine if stress changes in the 1989 earthquake moved it and segment 3 closer or farther from failure for the next large shock.

I presented what is given as figure 8.3 here and some of its implications to the Committee on Seismology of the National Academy Sciences in late 1986, and it was well received. I then sent our recommendations to Frank Press, the president of the National Academy of Sciences, who recommended that NEPEC should review the chances of a great earthquake in southern California.

Peck and Filson of USGS and I agreed to set up a separate working group to review those findings. Lloyd Cluff, formerly of Woodward-Clyde Associates, had just moved to Pacific Gas and Electric in San Francisco and said he was willing to be one of the chairs of the review committee if it would also examine northern California, especially the Hayward Fault. He and James Dietrich of USGS then chaired the working group, which included both northern and southern California. Their work was reviewed by NEPEC and was published in 1988 as the first Working Group Report on California Earthquake Prediction (USGS Working Group 1988). Another working group was established to review probabilities for the greater San Francisco Bay Area after the earthquake of 1989. Similar groups have continued to examine those areas of California about every five years.

Prediction of an Earthquake at Parkfield by the USGS

When I was first became chair of NEPEC, John Filson of USGS, who was also vice chair of NEPEC, asked us to evaluate an earthquake prediction for Parkfield in central California made by William Bakun and Allan Lindh of USGS in 1985. NEPEC then reviewed Parkfield at several meetings. As well as examining their prediction, we also reviewed a detailed response plan by USGS that indicated what should be done and what would be announced publically in the event various phenomena thought to be precursory occurred at Parkfield.

In retrospect, the most controversial aspect of the Bakun and Lindh prediction was their argument that the Parkfield earthquake of 1934 had occurred too early for a variety of reasons—that is, only 12 years after the previous shock in 1922. This led them to predict a 95 percent probability that the next earthquake at Parkfield after the one in 1966 would occur by 1993. The next earthquake did not occur, however, until 2004. Its magnitude and location were predicted correctly but not its timing. In retrospect, NEPEC should have examined more carefully the assumptions about the presumed early triggering of the 1934 shock and insisted that a probability calculation be made excluding those assumptions.

Bakun and Lindh calculated a very small uncertainty (standard deviation) of 1.6 years for the time intervals between Parkfield earthquakes. Nevertheless, if their assumption about the 1934 shock being triggered early is omitted, the standard deviation becomes much larger, 7.2 years. If probability calculations were made in 1985 using 7.2 years instead of 1.6 years, the 95 percent confidence limits would have extended into the earliest twenty-first century. Even then, however, the Parkfield earthquake of 2004 still occurred several years later than expected at that high level of confidence.

Parkfield events had been taken to be very regular, but after the 2004 shock it became clear that they were more irregular. In 2006, William Menke of Lamont and I calculated that the time intervals between Parkfield events, including the earthquake of 2004, were less regular than multiple large events along various fault segments in Alaska, Japan, Turkey, and elsewhere in California.

Why were Parkfield earthquakes more irregular in their timing? One possibility is that various Parkfield shocks did not rupture exactly the same segment of the San Andreas Fault. We know very little about events in 1881 and 1901 and how much the Parkfield segment broke in conjunction with the adjacent great earthquake to the southeast in 1857.

A likely explanation of this greater irregularity is that the Parkfield segment is the site of relatively small earthquakes, magnitude 6.0, compared to the less frequent but much larger events just to the southeast. Great earthquakes such as the one in 1857 may well influence the timing and size of Parkfield shocks.

The failure of the Parkfield prediction led to a major de-emphasis on earthquake prediction on all time scales by the USGS and a switch to major funding for estimating earthquake hazards and shaking. The USGS and the United States had put all of its eggs in the Parkfield basket even though several of us had recommended that more areas needed to be monitored and studied aggressively.

Soviet Prediction of a Large Earthquake in California and Western Nevada

In December 1986, Leon Knopoff, professor of geophysics at UCLA, sent me a letter asking NEPEC to review a prediction by V. I. Keilis-Borok and his colleagues of an earthquake of magnitude 7.5 or larger somewhere in California and western Nevada, derived at through the use of pattern-recognition techniques. Keilis-Borok, a full member of the Soviet Academy of Sciences, had delivered his prediction to Secretary-General Mikhail Gorbachev of the Soviet Union, who then had conveyed it to President Ronald Reagan at their summit meeting in Iceland in 1986. At USGS and NEPEC, we decided we could neither ignore that prediction nor Knopoff's request.

Many of us on NEPEC knew Keilis-Borok's work. His prediction, using a program called M8, was for a magnitude 7.5 or larger earthquake in a huge area, approximately 480 miles (775 kilometers) in diameter, between the beginning of 1984 and the first day of 1988. That area, centered just south of Yosemite National Park, included San Francisco, part of Los Angeles, and western Nevada. In a letter to me, Keilis-Borok said the prediction was for five years ending on December 31, 1988. He later extended it to 1991.

The Soviet computer program or algorithm M8 used a number of traits established earlier by pattern-recognition methods and applied them to catalogs of moderate-size earthquakes for California. Those traits were used to generate what the Soviet workers called Times of Increased Probability (TIPs).

Most of the earthquakes the Soviet group used to calibrate M8 involved thrust faulting at subduction zones. Few California earthquakes since 1906 were as large as magnitude 7.5. Hence, the Soviet predictions for California had to rely on an algorithm that was formulated for areas—that is, subduction zones—whose tectonic setting was quite different.

On April 1, 1987, NEPEC discussed the Soviets' prediction and a paper that described their method. We decided that we needed more information. Keilis-Borok then spent a day talking to us at the last meeting of NEPEC that I chaired in June 1988. At our invitation, two additional U.S. geophysicists attended the meeting and then submitted written comments.

In his letter to me in December 1986, Knopoff said that the methodology had been applied "to other recent large earthquakes with huge success." Many of us who were present at the NEPEC meetings would not have described these predictions as hugely successful. In our report in 1988, we said that the Soviets' prediction for California was of uncertain robustness and that its forecast area was very large. We concluded that the results did not warrant any special public-policy actions in California and western Nevada. All of the members of NEPEC concurred with that recommendation.

In our written report, we said, "The present work documents a maturing of the pattern-recognition approach to the analysis of global seismicity as a possible predictive tool for long-term earthquake prediction." We considered it a serious analysis of patterns of seismicity for earthquake prediction. We also stated, "The conclusions about the potential for a large earthquake in the California–Nevada region are intriguing, but many uncertainties about the approach and conclusions exist."

One of the problems with M8 was that the Soviet workers frequently extended predictions by one or more times when the same TIP indicated continuing anomalous behavior. When assessing the capabilities of M8, it is difficult to evaluate whether the initial period of alarm and its extensions should be counted as separate predictions. In addition, the Soviet method of prediction for California and western Nevada was only about a factor or two better than assuming random occurrence of large earthquakes.

Bernard Minster of UC San Diego, one of NEPEC's geophysical consultants, concluded that one of the weak points of the Soviets' prediction was that they dealt with such a small number of large earthquakes in California, so that it was difficult to generate statistical tests of the method. He said the Soviet report was difficult to read and to understand at a level of detail sufficient to permit easy duplication of the results.

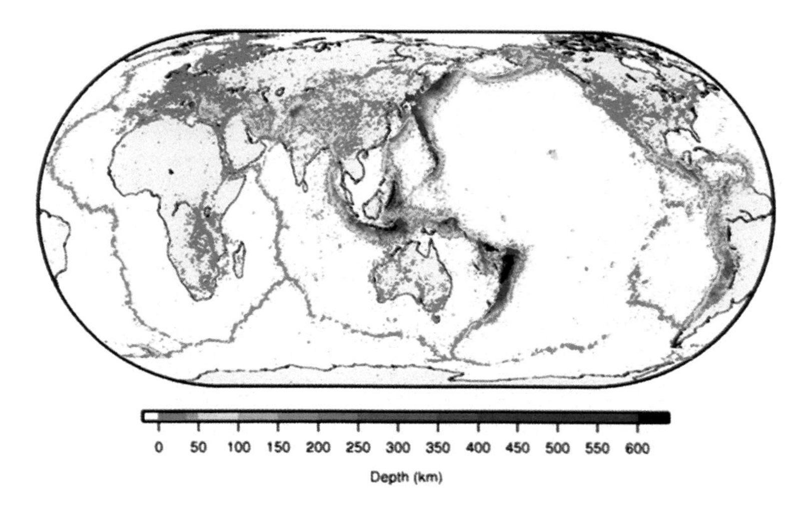

PLATE 1 Global locations of earthquakes from 1960 to 2014. Depths of events are color-coded.

Source: From the homepage of the *Bulletin of the International Seismological Centre* website, 2014, http://www.isc.ac.uk, updated once a year.

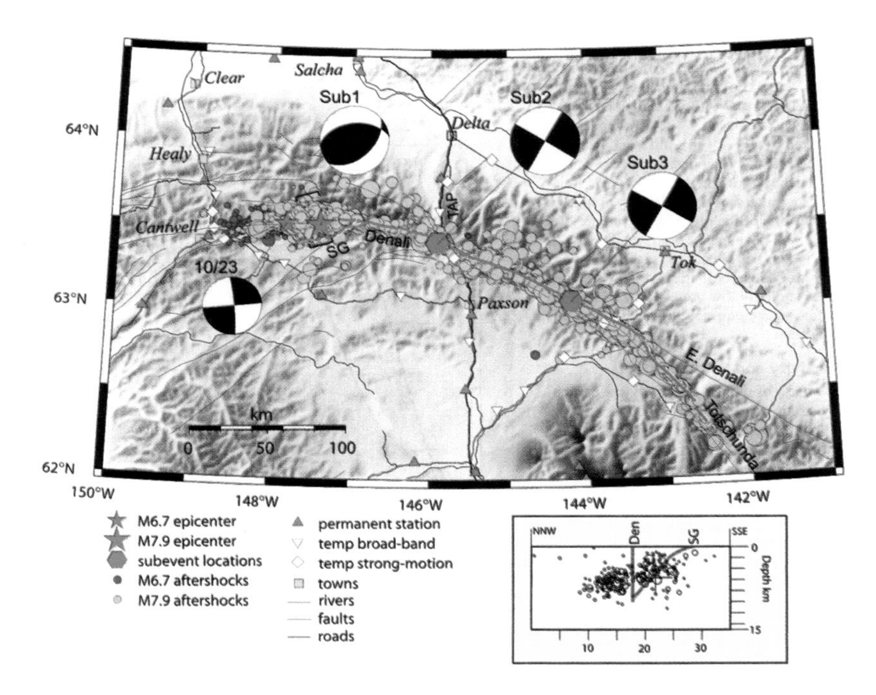

PLATE 2 Orange circles denote aftershocks of a great earthquake along the Denali Fault on November 3, 2002. Its magnitude, 7.9, was comparable to that of San Francisco earthquake of 1906.

Source: Eberhart-Phillips et al. 2003, with permission of the American Association for the Advancement of Science.

PLATE 3 Offset of a highway in central Alaska during the great earthquake of 2002 along the Denali fault. The Trans-Alaskan Pipeline crosses the fault near this location. Robert Page Jr. of the USGS insisted that the pipeline be built to accommodate such an offset, which it was.

Source: Photograph courtesy of Lloyd Cluff.

PLATE 4 The author taking a core of a tree along the Denali fault in 1967.

Source: Photograph courtesy of Robert Page Jr.

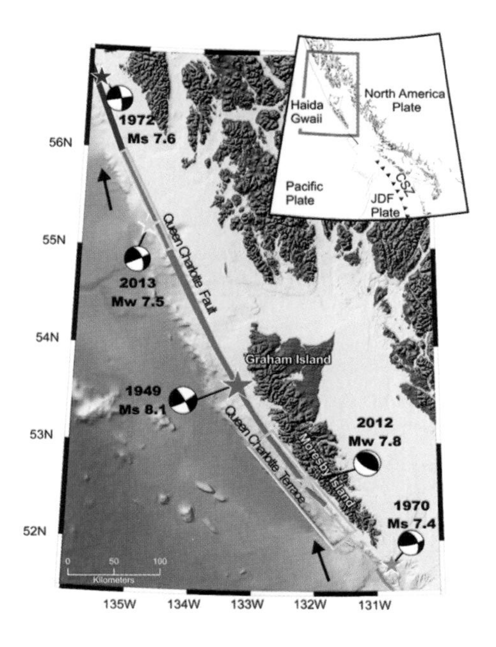

PLATE 5 Map of Haida Gwaii, British Columbia, and Craig, Alaska, earthquakes of magnitudes 7.8 and 7.5 in 2012 and 2013. The latter event filled a seismic gap south of the earthquake rupture zone in Alaska in 1972.

Source: James et al. 2015.

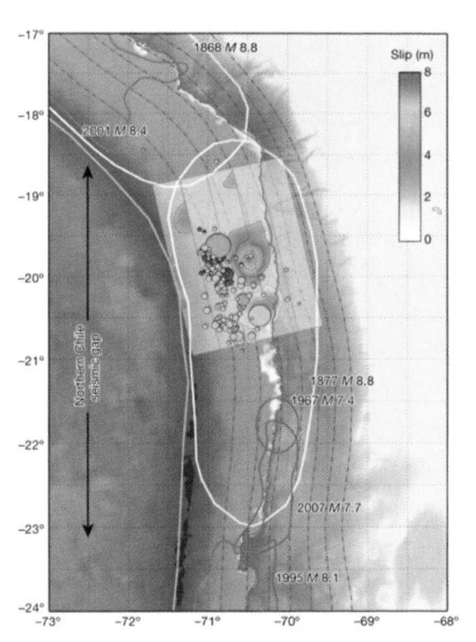

PLATE 6 Contours of slip in the great earthquake of April 1, 2014, offshore of Iquique, northern Chile. Rupture zones of other large earthquakes since 1900 are outlined in light purple. White lines enclose the inferred rupture zones of the giant shocks of 1868 and 1877. The large orange circle indicates epicenter of the 2014 mainshock; red circles denote its foreshocks; smaller orange circles indicate events between the mainshock and the largest aftershock of magnitude 7.7 (large yellow circle); small yellow circles indicate subsequent aftershocks.

Source: Hayes et al. 2014.

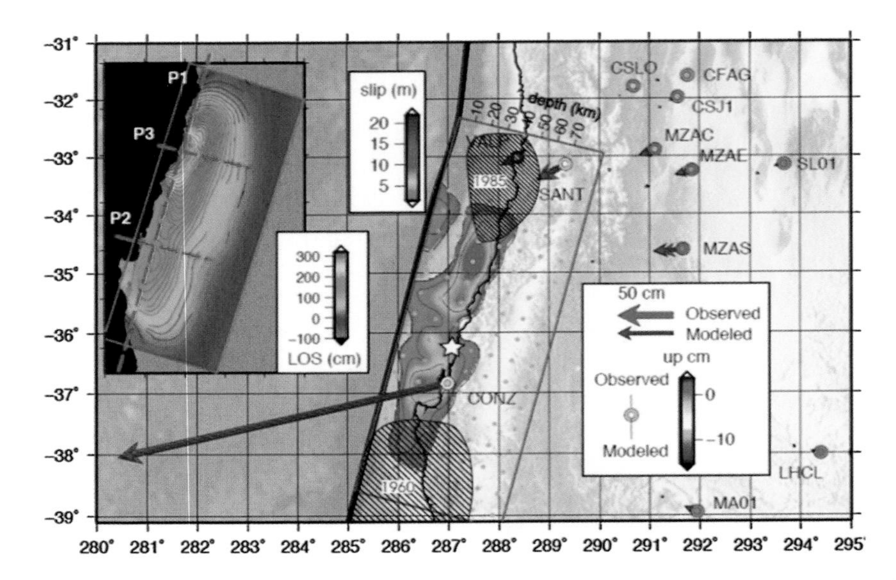

PLATE 7 Slip during giant Maule earthquake of 2010 in central Chile contoured in colors from larger than 23 feet (7 meters) in various shades of red and orange to about 3 feet (one meter) in blue. The white star denotes the epicenter, the point of initial rupture. The orange dots and arrows indicate amounts of slip at GPS stations. Hatched areas denote the rupture zones of the 1960 and 1985 earthquakes. The heavy black line is the axis of the Peru–Chile trench.

Source: Delouis, Nocquet, and Vallée 2010.

PLATE 8 View of the San Andreas Fault to the north of San Francisco (segment 2) looking eastward from the Pacific Ocean (*bottom*) to snow-capped Sierra Nevada peaks (*top*). Point Reyes Peninsula is at the bottom. Blue arrows point to the San Andreas Fault, which extends horizontally across the figure.

Source: Photograph courtesy of USGS, EROS Data Center, n.d.

PLATE 9 The San Andreas Fault at Point Reyes Station where a fence was offset during the 1906 earthquake. The fence has been restored.

Source: Photograph of the author taken in 2006.

PLATE 10 Collapse of the Cypress Street Viaduct on the east side of San Francisco Bay during the earthquake in 1989.

Source: Photograph courtesy of Lloyd Cluff.

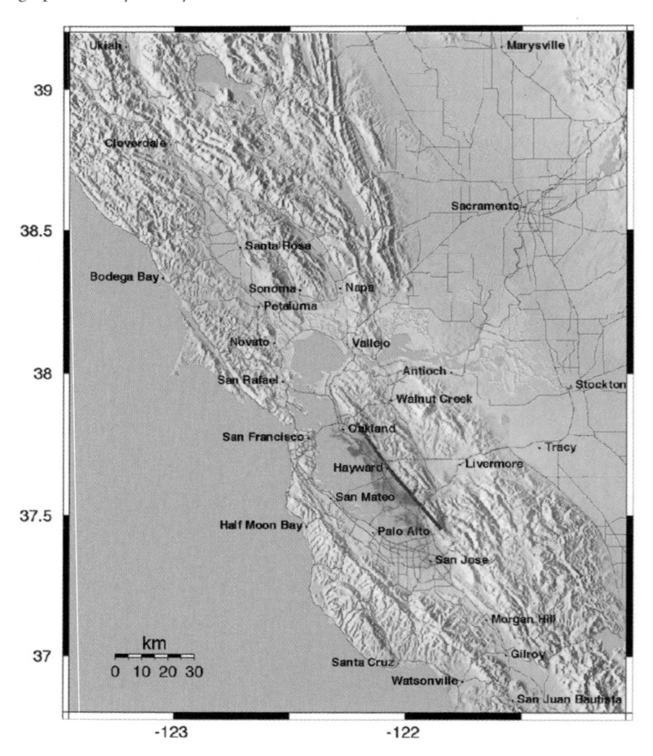

PLATE 11 Projected levels of shaking in a future earthquake of magnitude 7.0 that would rupture the entire Hayward Fault (*heavy black line*). Red colors indicate strongest shaking.

Source: USGS 2016.

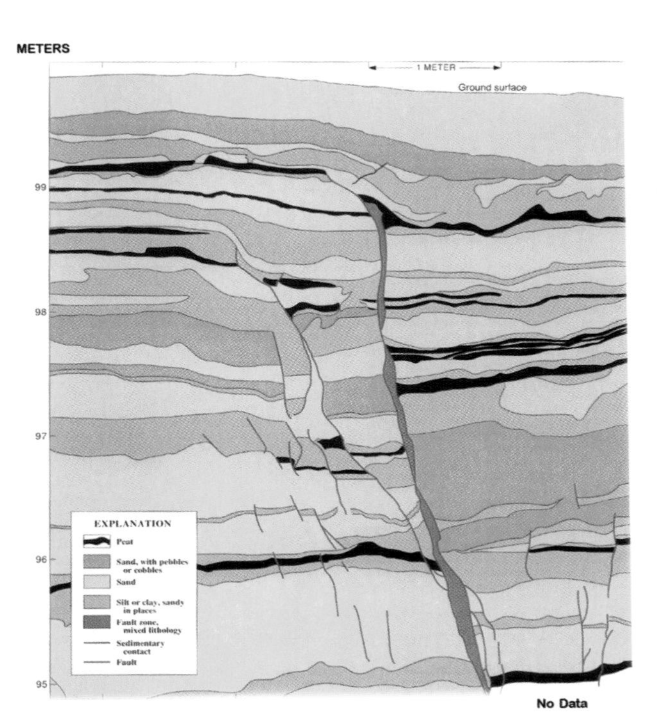

PLATE 12 The San Andreas Fault exposed in the southeast-facing wall of a trench excavated by Kerry Sieh in the 1970s at Pallett Creek. The fault zone is indicated by the purple color; individual fault breaks in three past earthquakes are indicated by red lines. The excavation extends from the surface to a depth of about 16 feet (5 meters). The Pacific plate is to the left, and the North American plate is to the right.

Source: USGS 1990b.

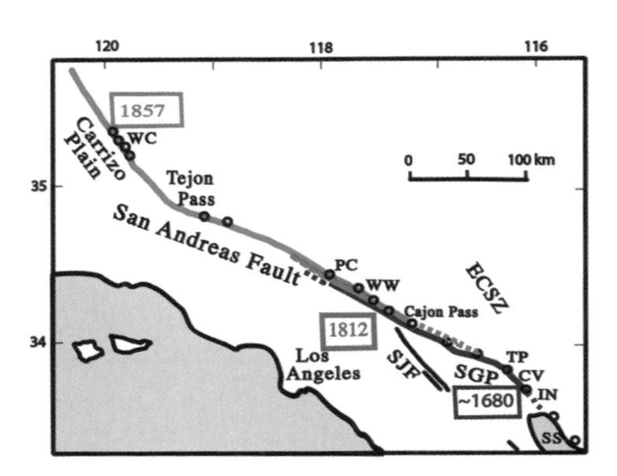

PLATE 13 Parts of the San Andreas Fault that ruptured in the great earthquakes of 1857, 1812, and around 1680 in southern California. PC is the site of the trench at Pallett Creek shown in figure 8.13. Open black circles indicate paleoseismic (trenching) sites.

Source: Sykes and Menke 2006.

PLATE 14 Northeast near the town of Midori, Japan, bushes in a field mark an old property line that was offset to the left about 10 feet (3 meters) during the earthquake in 1891. The old road at the left side of the field also was offset to the left to the position marked by the seismologist standing on the rocks. The fault extends from the center of the right side of the photo to the seismologist's feet. The fault motion in 1891 also uplifted the road about 7 feet (2 meters).

Source: Photograph by the author, 1974.

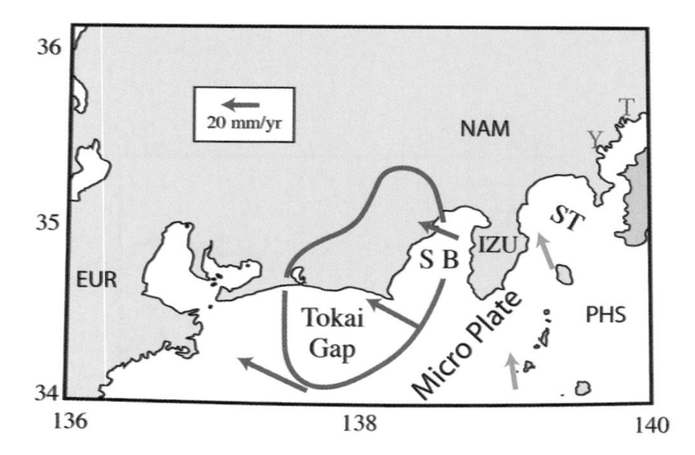

PLATE 15 Suruga Bay (SB), the Izu Peninsula (IZU), Tokyo (T), and Yokohama (Y) at the far upper right side of the figure. The great earthquake of 1923 occurred along the Sagami Trough (ST). The red oval encloses the Tokai seismic gap, which has not broken in a large earthquake since 1854. EUR is the Eurasian plate; PHS, the Philippine Sea plate; NAM, the North American plate. Red and blue arrows indicate plate convergence.

Source: Sykes and Menke 2006.

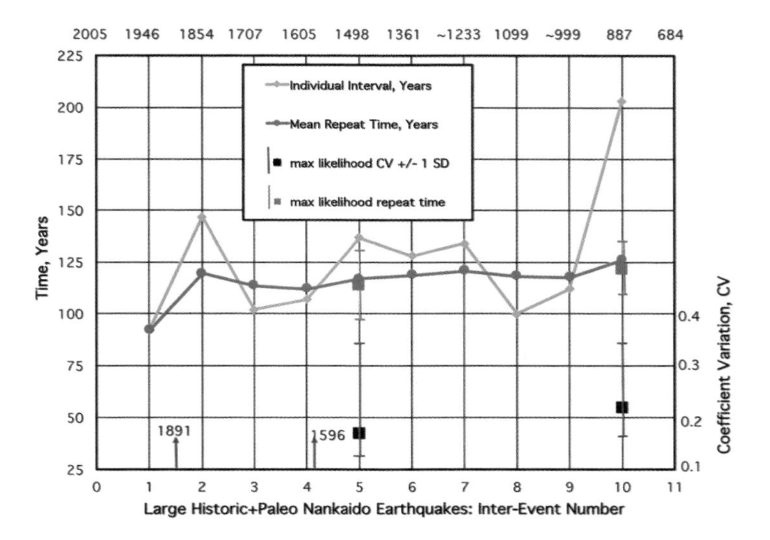

PLATE 16 Repeat times of great earthquakes along segment B (figure 10.3) of the Nankai plate boundary. The two arrows at the bottom indicate the great inland earthquakes of 1596 and 1891, which were not located along the plate boundary.

Source: After Sykes and Menke 2006.

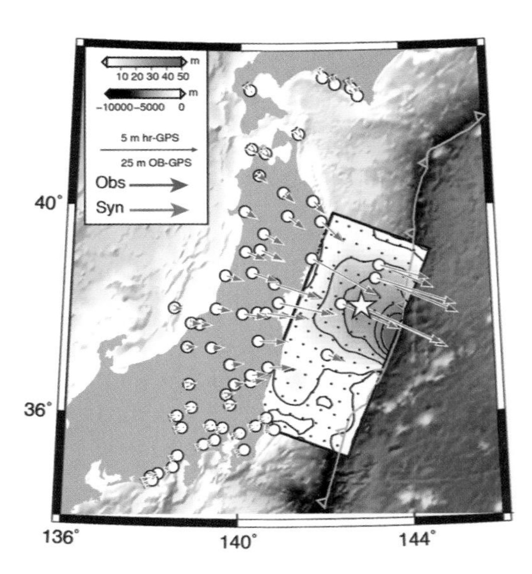

PLATE 17 Fault displacements (horizontal slip) during the Tohoku mainshock of March 11, 2011. The large open star denotes its epicenter. The light-blue arrows denote GPS observations of slip; the red arrows indicate calculated slip. The solid black box encloses the approximate rupture area of the mainshock. Red colors of increasing intensity indicate greater amounts of slip. The blue-black color indicates the deepest part of the Japan Trench, the easternmost extent of slip.

Source: Yue and Lay 2013.

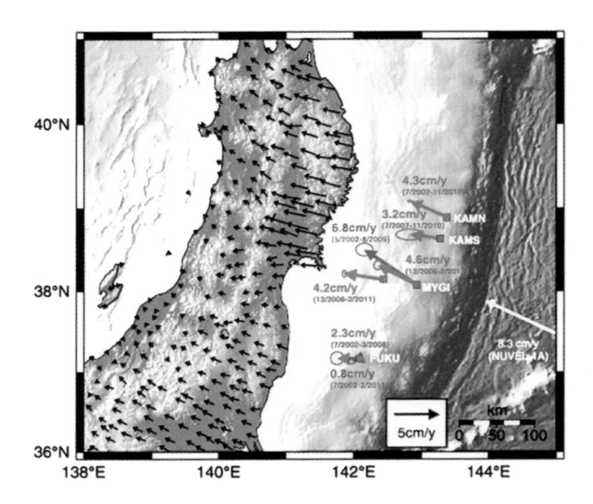

PLATE 18 Horizontal displacements per year on land (black arrows) and at five stations on the seafloor prior to the 2011 earthquake. The white arrow at the lower right indicates the long-term rate of motion of the Pacific plate with respect to Honshu, 3.3 inches per year (8.3 centimeters per year). Note that point MYGI moved northwesterly at 82 percent of the plate rate between 2002 and 2005.

Source: After Sato et al. 2013.

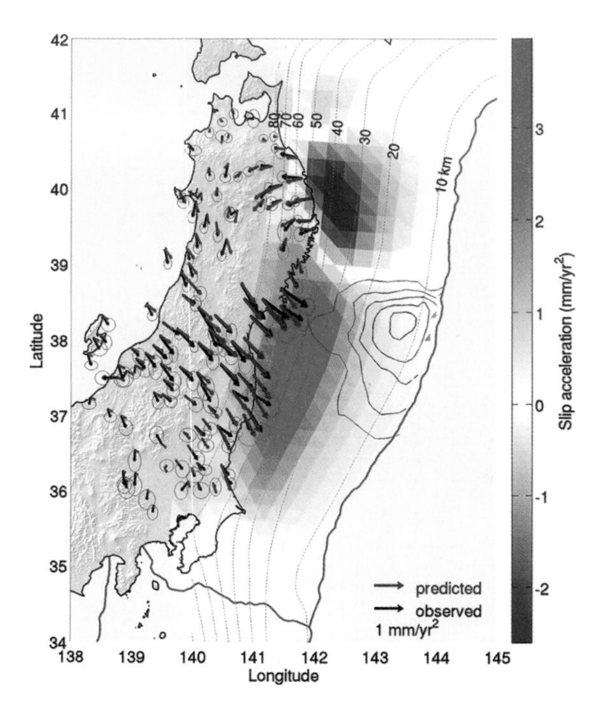

PLATE 19 Slip on the plate boundary as deduced from GPS stations on land from 1996 until just before the giant earthquake of 2011. Slip accelerated in pink and red areas and decelerated in blue areas.

Source: Mavrommatis, Segall, and Johnson 2014.

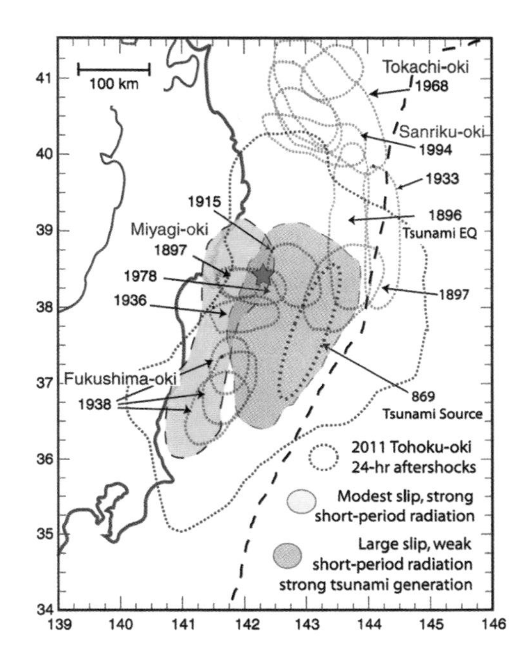

PLATE 20 Rupture areas of the Tohoku–Oki earthquake in 2011 that generated strong, short-period seismic waves (*lighter shade*) and waves with large slip but weak short-period radiation (*darker shade*) closer to the Japan Trench. The thick dashed black line indicates the deepest part of the trench. The star denotes the epicenter.

Source: Koper et al. 2011.

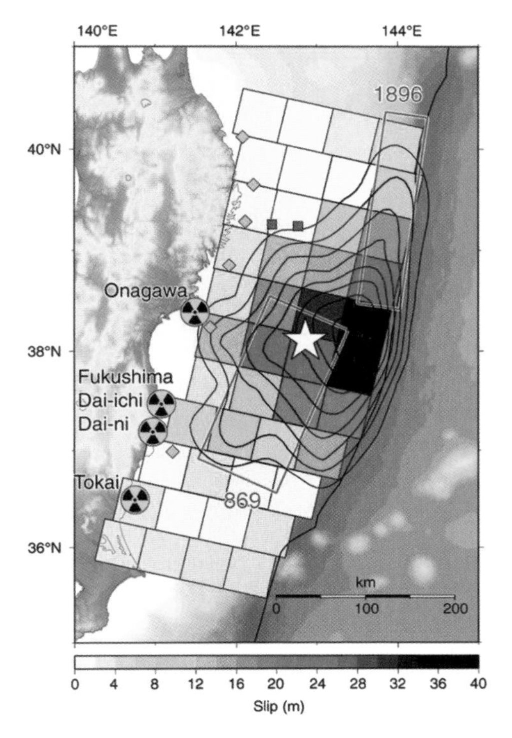

PLATE 21 Cumulative slip in the giant earthquake of 2011; the large star indicates the epicenter. Radiation symbols mark locations of four nuclear-power plants. The outer blue line is the deepest part of Japan Trench.

Source: Satake et al. 2013.

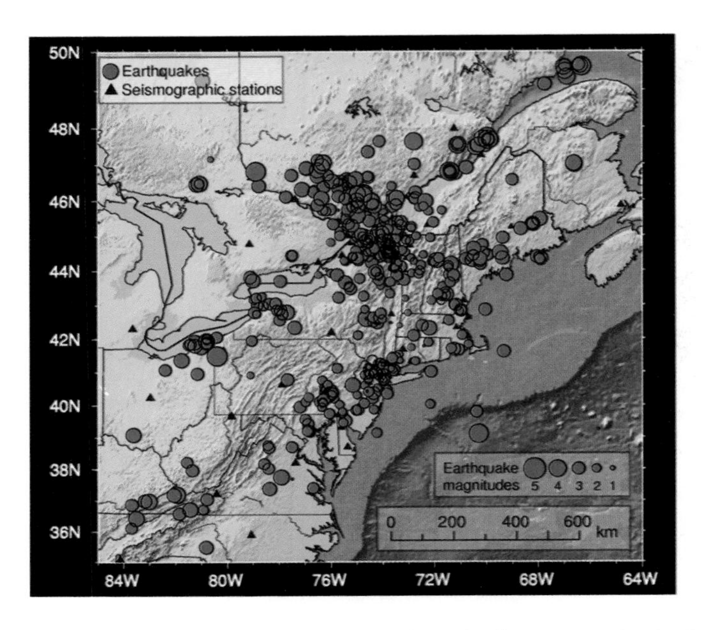

PLATE 22 Earthquakes of magnitude 2.5 and greater in the mid-Atlantic states, New England, and adjacent parts of Canada from 1977 to 1999. Blue circles denote shocks of magnitude greater than 4.0.

Source: Lamont-Doherty Earth Observatory unpublished report, 2003.

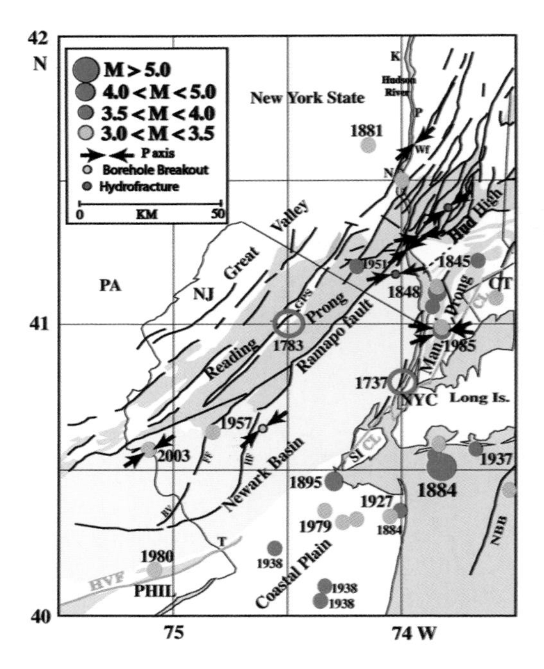

PLATE 23 Earthquakes in the greater New York City–Philadelphia area, 1700 to 2004.

Source: Sykes et al. 2008.

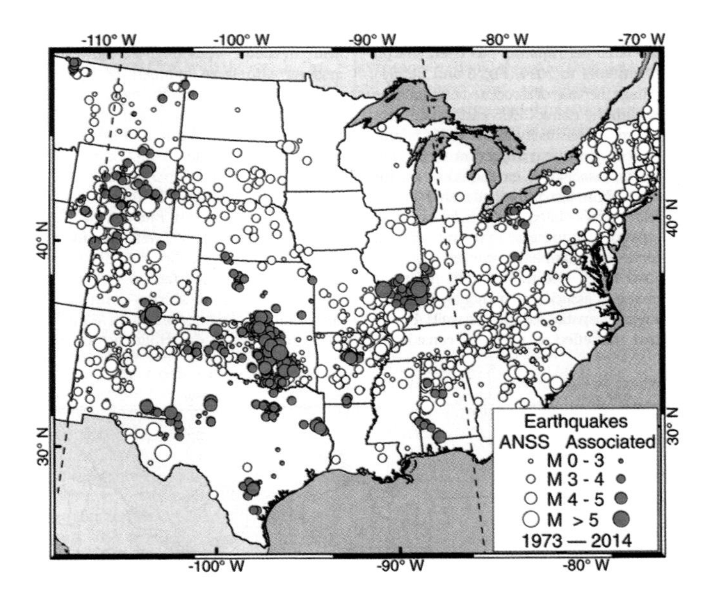

PLATE 24 Earthquakes within the North American plate since 1700.

Source: Weingarten et al. 2015.

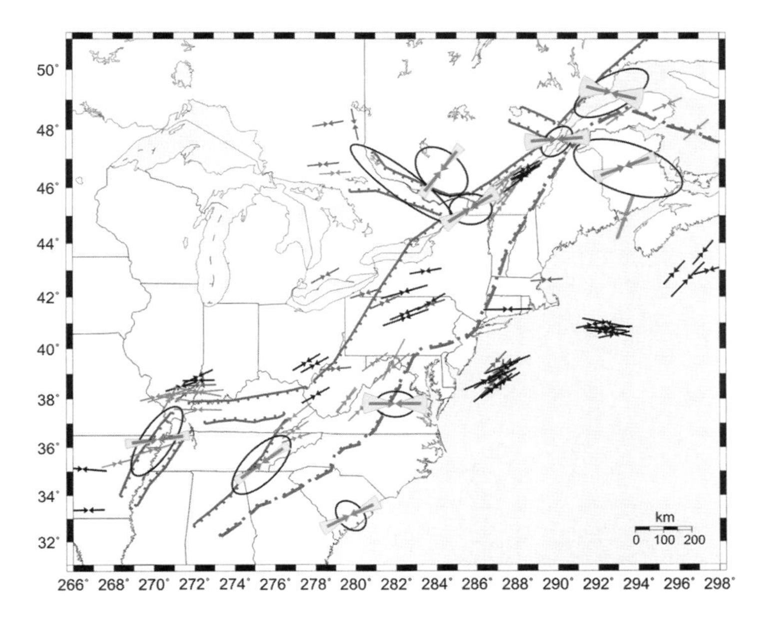

PLATE 25 Directions of compressive stress in the eastern United States and adjacent parts of Canada. Red arrows indicate directions from earthquake mechanisms. Arrows of other colors denote measurements in boreholes.

Source: Mazzotti and Townsend 2010.

PLATE 26 Indian Point nuclear-power plants, Buchanan, New York, looking east across the Hudson River.

Source: Photograph courtesy of Tony Fisher, 2008, Wikimedia Commons.

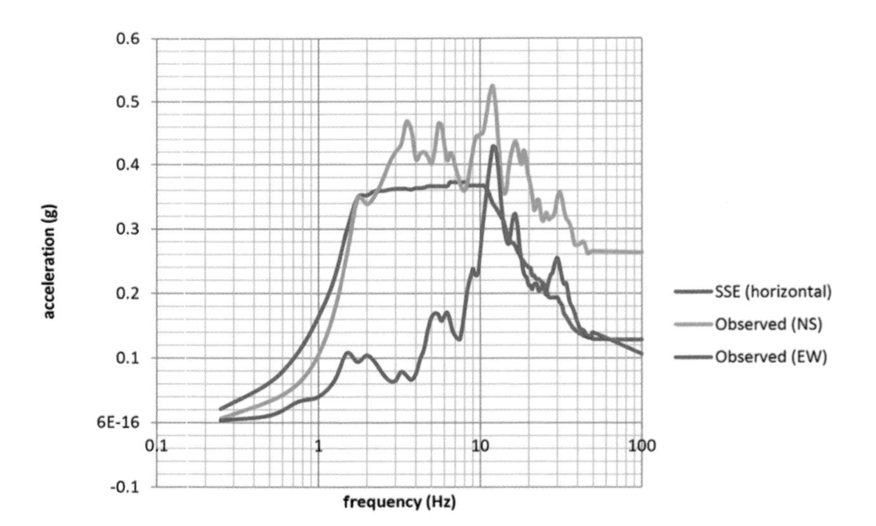

PLATE 27 Comparison of observed horizontal accelerations of the surface of the earth during the Virginia earthquake of 2011 as a function of frequency. Calculated values for the safe-shutdown earthquake (SSE) at North Anna are indicated in red.

Source: U.S. NRC 2011.

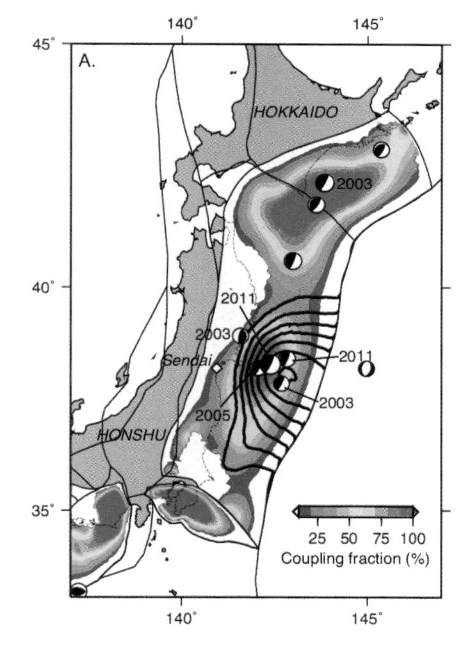

PLATE 28 Strong seismic coupling along several parts of plate boundaries off Japan as determined from GPS observations.

Source: Loveless and Meade 2015.

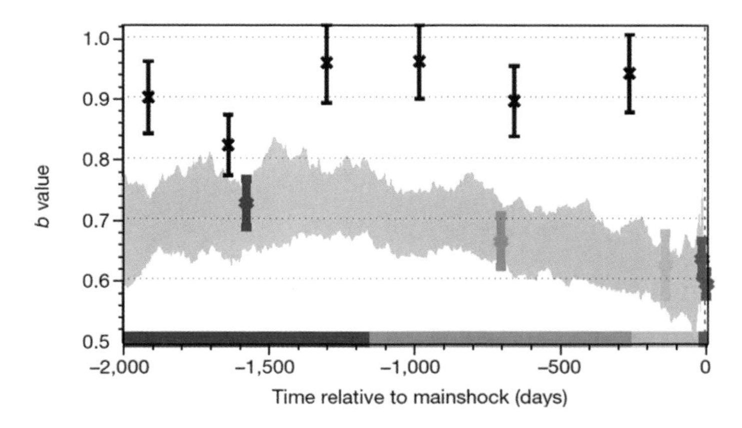

PLATE 29 Decrease in *b* values in coming rupture zone 5.5 years before Iquique mainshock of 2014 (*colored symbols and purple area*). Black bars indicate higher, normal values in adjacent areas outside of that rupture zone.

Source: After Schurr et al. 2014.

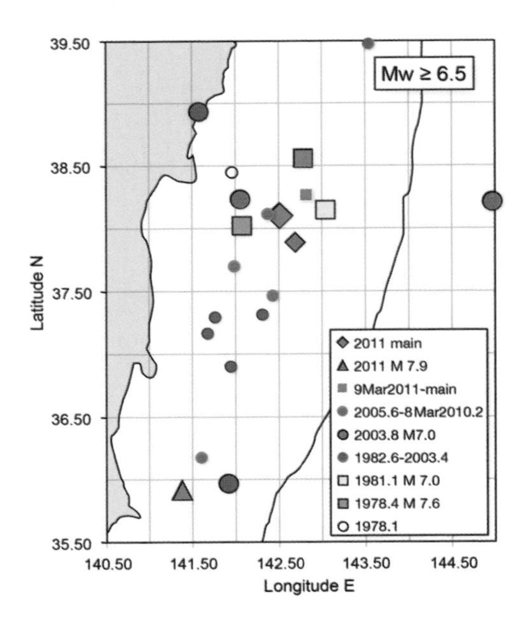

PLATE 30 Earthquakes of magnitude 6.5 and larger from 1976 through the occurrence of the giant earthquake on March 11, 2011, and its largest aftershock of magnitude 7.9. Larger symbols denote earthquakes of magnitude 7.0 and greater.

Source: Unpublished figure by the author, 2016.

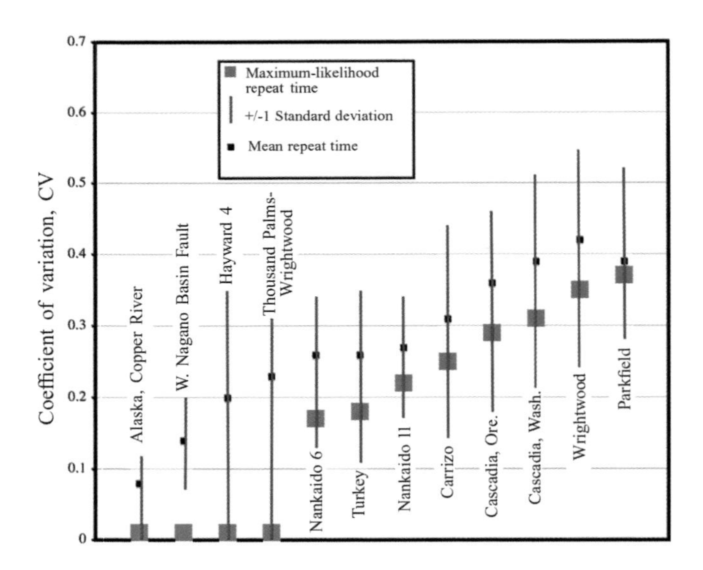

PLATE 31 Coefficient of variation, CV, a measure of the uncertainty in individual time intervals between large shocks that have ruptured various segments of active plate boundaries.

Source: Sykes and Menke 2006.

Keilis-Borok was pleased with what he considered to be constructive comments made at our meeting but not with our recommendations to the USGS director, which we also sent to Keilis-Borok. The Soviet group continued to make predictions of large earthquakes in California, none of which have occurred as of late 2018.

Opposition to NEPEC's Wider Role

Soon after our first meeting in 1984, I received a letter from NEPEC member Thomas McEvilly of UC Berkeley cautioning that NEPEC should be careful in taking a more activist role in examining the likelihood that some regions of the United States could be the sites of large and damaging earthquakes on time scales of a few decades. I had recommended that role, and it was approved by the USGS director and by NEPEC itself. McEvilly stated, "Is our advisory role to the [USGS] Director compromised by our hands-on role in his research program?" He considered NEPEC's first order of business to serve as an advisory council that would be called upon in situations of some emergency when a prediction must be evaluated and generally a public statement issued. I disagreed then and still do with that description of NEPEC's main role.

When I left NEPEC in the summer of 1988, USGS asked McEvilly to chair the council. As far as I can see, NEPEC has now taken a passive approach of hearing only predictions that are brought to it, as McEvilly advocated. I presume that was USGS's choice as well, given the failure of its Parkfield prediction. USGS decreased funding for earthquake studies and put greater concentration on estimating shaking, damage, and loss of life in future U.S. earthquakes. During the four years that I was NEPEC's chair, we addressed the likelihood that some regions and faults might or might not be the sites of large and damaging earthquakes on time scales of a few decades. I think it was a valuable endeavor that is largely overlooked today in the United States.

(I have a thick folder of unpublished NEPEC documents, which will be preserved and available for future research at the Rare Book and Manuscript Library of Columbia University.)

10

JAPANESE EARTHQUAKES AND THE FUKUSHIMA NUCLEAR DISASTER

Five giant earthquakes occurred worldwide during the decade 2004 to 2013, more than in any ten-year period during the past century. These events were especially well recorded and analyzed by modern seismological, geodetic, and ocean-bottom techniques. Observations and studies of them provide new insights into the generation of earthquakes and possibilities for long-term prediction.

One of these events—the giant Tohoku earthquake of 2011 and its accompanying tsunami—triggered the Fukushima nuclear disaster. In this chapter, I focus on several significant, damaging, and great Japanese earthquakes to illustrate what they reveal about the physics of earthquakes, long-term prediction, the subduction process, and the generation of tsunamis.

All but two of the giant earthquakes from 1900 to 2015 occurred at subduction zones. Several of them, such as the Tohoku earthquake of 2011, generated large and destructive tsunamis. One giant shock involved underthrusting of the Indian continental crust beneath Tibet along the eastern Himalayas in 1950. Like the giant shocks at subduction zones, this shock involved horizontal compression and convergence of two tectonic plates along gently dipping plate interfaces. A single different giant shock occurred west of Sumatra in oceanic crust in 2012 and is the largest known strike-slip earthquake.

Mino-Owari 1891

The Mino-Owari (Nobi) earthquake of 1891, which occurred within the main Japanese island of Honshu to the north of the city of Nagoya, ushered in the era of modern seismology. It killed more than 7,200 people and destroyed more than 17,000 buildings. Some of the first seismograph records were made of this event, indicating its magnitude was about 7.5. After it occurred, the Japanese government formed an earthquake-investigation committee and funded geodetic surveys to measure horizontal and vertical deformations related to earthquakes.

Because the earthquake in 1891 occurred on land and not at sea, as in many large subduction-zone earthquakes off the coasts of Japan, the geologist Bunjiro Kotō was able to map fault rupture at the surface. He described those breaks as an "earth rent" resembling the path of a giant mole. The breaks consisted of a series of closely spaced faults extending northwesterly for about 70 miles (112 kilometers) along the Neo Valley. The California earthquake of 1906 generated a similar "mole track" (figure 8.8). Strike-slip faulting was reported for the shock in 1891, 15 years before it was recognized along the San Andreas Fault.

Kotō concluded that sudden displacements along the Neo Valley fault were the actual cause of the earthquake. Nevertheless, until around 1960 many geophysicists in Japan considered earthquakes to be a secondary phenomenon, a form of damage resulting from primary events, such as volcanism. This conclusion was understandable in Japan because volcanoes are numerous there and most large earthquakes occur at sea, where, until recently, surface rupture could not be observed directly.

Similarly, about 50 years ago New Zealand geophysicist Frank Evison was one of the last to propose that earthquakes are a secondary consequence of some other more fundamental earth processes. Since then, earth scientists in Japanese and New Zealand, like others worldwide, have accepted that earthquakes are generated directly by sudden slip along a fault and by a drop in stress in a volume of rock surrounding that fault segment.

In his book *Elementary Seismology* (1958), Charles Richter reproduced a famous photograph Kotō took in 1891 of faulting in fields near the town of Midori. The earthquake offset many small agricultural plots of land. Instead

FIGURE 10.1 Northeast near the town of Midori, Japan, bushes in a field mark an old property line that was offset to the left about 10 feet (3 meters) during the earthquake in 1891. The old road at the left side of the field also was offset to the left to the position marked by the seismologist standing on the rocks. The fault extends from the center of the right side of the photo to the seismologist's feet. The fault motion in 1891 also uplifted the road about 7 feet (2 meters). See plate 14.

Source: Photograph by author, 1974.

of straightening the boundaries of their plots after the earthquake, farmers kept the now irregular boundaries, preserving the offset seen today by patterns of different crops and brush. My photo of the fault offset (figure 10.1), which I took in 1974 near Midori, still clearly shows the horizontal and vertical displacements remaining from 1891.

Tokyo 1923

The magnitude 7.9 Kantō earthquake of September 1, 1923, created a huge disaster for Tokyo, Yokohama, and adjacent parts of the very populated

Kantō region. Official figures indicate that more than 99,000 people died, many in a firestorm triggered by the great earthquake. The 2010 census shows that the population of the Kantō region, which includes Tokyo, increased greatly after 1923 to around 42 million, about one-third of the population of Japan. Hence, many more people are exposed to the risk of a future great shock today.

The mechanism of the earthquake in 1923 involved a combination of thrust and strike-slip faulting along the plate boundary to the south of Tokyo (T in figure 10.2). This disaster led to the establishment of the Earthquake Research Institute of Tokyo University.

Frank Lloyd Wright's famous Imperial Hotel officially opened on the day of the earthquake. Designed with earthquake-resistant features, it survived the shock with only modest damage. Its reflecting pool provided water to fight the subsequent firestorm. According to reports, however, about 19 percent of brick buildings and 20 percent of steel and reinforced concrete buildings in Tokyo performed better than the Imperial Hotel. The hotel's main failing was its poor foundation on soft sediments. Much of Tokyo and the shores of Tokyo Bay sit on similar poor soils, which amplify seismic waves of certain frequencies.

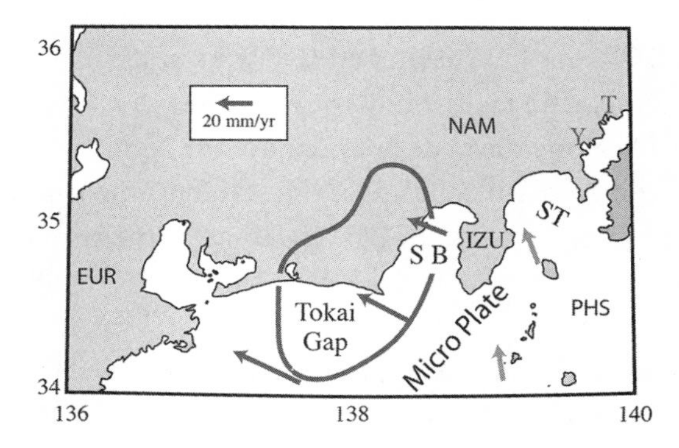

FIGURE 10.2 Suruga Bay (SB), the Izu Peninsula (IZU), Tokyo (T), and Yokohama (Y) at the far upper right side of the figure. The great earthquake of 1923 occurred along the Sagami Trough (ST). The oval encloses the Tokai seismic gap, which has not broken in a large earthquake since 1854. EUR is the Eurasian plate; PHS, the Philippine Sea plate; NAM, the North American plate. The arrows indicate plate convergence. See plate 15.

Source: Sykes and Menke 2006.

The 1923 earthquake ruptured the northwestern part of the boundary between the North American and Philippine Sea plates along the Sagami Trough to the south of Tokyo (figure 10.2). The rupture also extended inland along the north side of the Izu Peninsula. The peninsula, which is located on the Philippine Sea plate, collided with the rest of the island of Honshu sometime in the past 1 to 2 million years, a very short time geologically. Before that, the Izu Peninsula was a separate island on the Philippine Sea plate, like other islands of the Izu-Bonin subduction zone farther south. In a sense, the collision is like that of a "little India." (India moved northerly as a separate continent from a much more southerly location and then collided with the rest of Asia about 50 million years ago.)

A great earthquake in 1703 broke part of the zone that ruptured in 1923. The 1703 event also ruptured farther southeast along the Sagami Trough just four years before a giant earthquake broke the plate boundary along the Nankai Trough in 1707 (figure 10.3). Rupture in the Ansei One shock of 1854 extended into Suruga Bay (SB in figure 10.2), which is located to the northwest of the Izu Peninsula.

Great Earthquakes Along the Nankai Trough

The shock of 1707 and the Ansei I and II shocks of 1854 were typical great subduction-zone earthquakes. Similar earthquakes are well known historically back to 684 along the plate boundary of the Nankai subduction zone (figure 10.3). Segments A through E of the Nankai zone ruptured in single great or giant earthquakes, as in 1707 and 684. Some segments, however, ruptured separately but close in time, as in the 1944–1946, 1854, 1360–1361, and 1096–1099 pairs.

Nankai segment E (figure 10.3) appears to have ruptured less frequently in great earthquakes than segments A through D. Zone E did not break in1944 and has not ruptured in a great shock since 1854. In 1981, Japanese seismologist K. Ishibashi of Tokyo University proposed that a "soon-to-occur" great earthquake would rupture what became known as the "Tokai seismic gap," which is outlined in the large oval in figure 10.2. It includes the huge industrialized area to the southwest of Tokyo. The Japanese government built many high walls to protect the adjacent coastlines from a tsunami generated by a great earthquake in the Tokai gap.

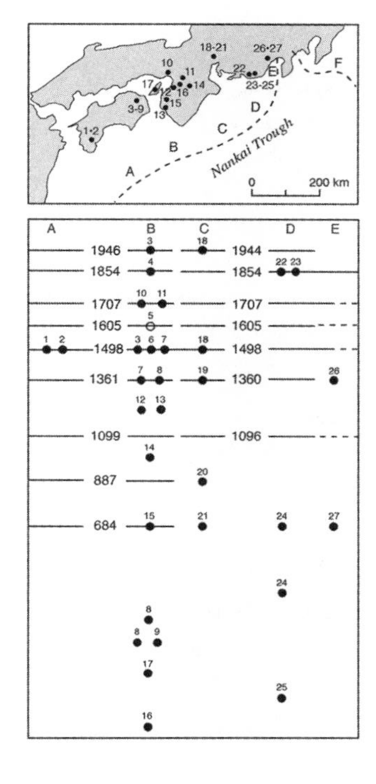

FIGURE 10.3 Historical earthquakes along segments A through E of the Nankai subduction zone. Dotted lines on the map (*top panel*) indicate the outcrop of the plate boundary at the seafloor. Solid horizontal lines (*bottom panel*) denote dates of historical shocks; dotted lines indicate questionable ruptures. Solid dots denote locations of events identified by A. Sangawa at the numbered archaeological sites. The great earthquakes of 1498 and 1891 were located inland and did not occur along the plate boundary.

Source: Sykes and Menke 2006.

Why has the Tokai earthquake forecast in 1981 not occurred yet? The rupture zones of great thrust earthquakes along segments A through D of the Nankai plate boundary are quite narrow and relatively simple. Their down-going plate interfaces are similar to the sketch in figure 10.4. In contrast, plate deformation along segment E includes not only Suruga Bay but also the entire width of the Izu microplate, as shown in figure 10.2. GPS data indicated by the arrows show that plate motion is spread across a broad zone. Hence, the long-term rate of plate motion across Suruga Bay is less than that along segments A through D to the southwest. This indicates that great earthquakes likely occur less frequently along Suruga Bay than along the simpler Nankai segments to the southwest.

Many past great earthquakes that broke segments A to D of the Nankai subduction zone in figure 10.3 may not have ruptured segment E located in Suruga Bay. Segment E and the Tokai seismic gap may not break in a great earthquake for many decades. It may instead rupture next in conjunction with a future great event along segment D, which last broke in 1944. Thus, the Tokai prediction was for a more complex region than other parts of the Nankai zone.

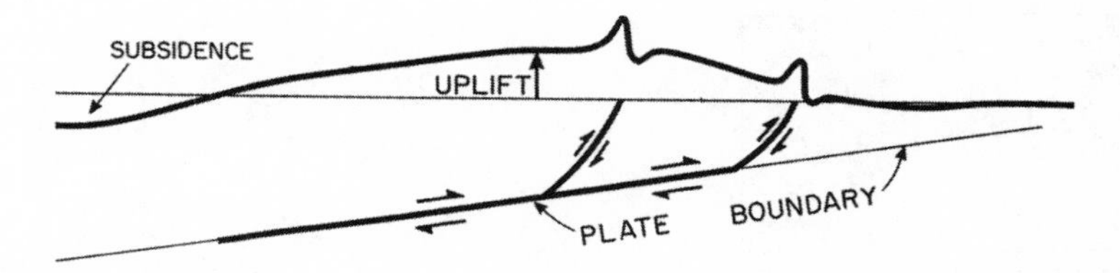

FIGURE 10.4 Simplified vertical cross-section of a subduction zone. Oceanic lithosphere on the right below the plate boundary underthrusts the upper plate on the left. Depth on the vertical axis extends from the seafloor to 45 miles (0–70 kilometers). The double arrows denote slip during a great earthquake. Vertical motion in that event is shown at the top.

Source: After Sykes and Menke 2006.

The English expression "soon to occur," first used by Ishibashi in 1981 about a future great earthquake in the Tokai gap, has also been used several times for other seismic gaps in Japan. The original Japanese text, however, does not appear to be more time specific than a few decades. Nevertheless, many in Japan unfortunately took "soon to occur" to imply a much shorter time span, expecting it would be the next great Japanese earthquake. With 36 years having elapsed as of mid-2018, the prediction of 1981 should now be considered to be a failure. Since 1981, other great earthquakes have taken place elsewhere in Japan.

Much can be learned from the Tokai example about how and how not to communicate predictions to government officials and the public. Legitimate concerns exist about the repeat of great earthquakes along segments A to E of the Nankai zone, especially as the 90-year interval between the shocks of 1854 and 1944 will be approaching in the few decades around 2034.

Repeat Times of Great Earthquakes Along the Nankai Subduction Zone

Improving long-term earthquake predictions necessitates using accurate information about the repeat times of past large and great earthquakes. Segments A to D of the Nankai plate boundary are well suited for this task

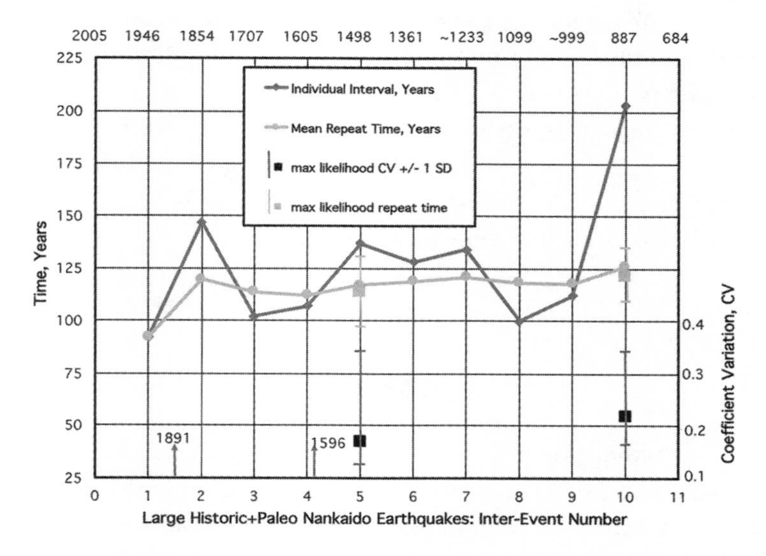

FIGURE 10.5 Repeat times of great earthquakes along segment B (figure 10.3) of the Nankai plate boundary. The two arrows at the bottom indicate great inland earthquakes of 1596 and 1891, which were not located along the plate boundary. See plate 16.

Source: After Sykes and Menke 2006.

because records of past great events extend back approximately 1,330 years, one of the longer historical records in the world.

The top of figure 10.5 indicates the dates of past great earthquakes along segment B of the Nankai plate boundary. The events are numbered from 1 to 11 going backward in time from the most recent great shock in 1946 because dates and repeat times since 1361 are more accurate than earlier ones. The average of the individual time intervals between great shocks since 1361 is about 114 years. The 203 years between the great shocks of 887 and 684 is anomalously long. William Menke of Lamont and I concluded in 2006 that the reason for this long period is that a great event is likely missing in the historical records between those two early shocks.

Because great earthquakes along Nankai segment B did not recur with exact periodicity, Menke and I wanted to examine *variations* in those repeat times. We calculated the date of the next great earthquake as 2060 ± 19 years—that is, between 2041 and 2079, with a probability of

68 percent. Remember, however, that the two most recent great events along segment B, those of 1946 and 1854, occurred just 92 years apart, one of the shortest repeat times of past shocks along the Nankai zone. If we used only that one repeat interval, the projected date of the next great shock would be 2038. Nevertheless, in making a long-term prediction for the next great shock, I advocate using longer series of time intervals between past earthquakes, such as those recorded since 1361.

Kobe 1995

The Kobe earthquake of 1995 in western Honshu caused the most damage in Japan between the great Tokyo event of 1923 and the giant shock of 2011. The amount of damage is surprising because the earthquake's magnitude was only 6.9. Nevertheless, the fault on which it occurred was close to the city of Kobe, population 1.5 million. Many structures in the central and port areas of the city that were situated on poor soils, including the Hanshen Expressway, were either destroyed or heavily damaged. Damaged structures also led to the closure of the main Shinkansen rail line (with its high-speed "bullet trains"). Major fires were widespread.

More than 6,400 people were killed in the Kobe earthquake, and losses totaled about U.S.$100 billion. After the Kobe disaster, the Japanese government allocated funds for more than 1,000 GPS stations and several modern arrays of seismic instruments. Their data proved to be very valuable during the 2011 earthquake. The very large financial losses in 1995 are typical in developed countries for large but not great earthquakes within urban areas. Similarly, the total monetary losses in the Northridge, California, earthquake of magnitude 6.6 in 1994, which occurred in the northern part of the Los Angeles metropolitan area, were about $55 billion. About half were covered by insurance.

The mechanism of the Kobe earthquake involved a combination of strike-slip and thrust faulting. It and many other earthquakes on land within western Honshu, such as those of 1891 and 1596, involved maximum compressive stress oriented east–west. The area is very active, although its earthquakes do not occur along a plate boundary.

Visits to Japan

I spent two months in Japan in late 1974 after visiting China for five weeks. I did much scientific and regular tourism throughout the country because it was my first visit. I took the train to Sendai in northeastern Honshu, which was struck later by a tsunami during the giant earthquake of 2011. Seismologists at Tohoku University were doing outstanding work locating earthquakes along their subduction zone. I visited Kyoto and Nagoya Universities on the bullet train from Tokyo. I was exhausted after many long days of visits to scientific labs and universities and long evenings of drinking with Japanese colleagues, all males. I took a long trip to the southern tip of the island of Kyushu to visit a very active volcano, Sakurajima. It welcomed me with several small explosions, a common occurrence. On my way back by train to Tokyo, I stopped for a day at Hiroshima. I returned to Japan two more times for meetings on earthquake prediction.

Giant Tohoku Earthquake and Tsunami of 2011

While vacationing in Key West, Florida, on March 11, 2011, I awakened to scenes on TV of a tsunami (seawave) inundating the airport and city of Sendai along the Pacific coast of Japan. Japan's largest-known earthquake in its long history of destructive shocks had generated this huge tsunami. The giant earthquake of magnitude 9.0 ruptured the subduction zone off northeastern Honshu, the main island of Japan. The maximum tsunami height was 133 feet (40.5 meters) with a wave that traveled up to 6 miles (10 kilometers) inland near Sendai, the major city of the region.

More than 18,000 lives were lost in the earthquake and tsunami. The World Bank estimated the economic costs as U.S.$235 billion, the costliest natural disaster in world history. More than 127,000 buildings totally collapsed, and an additional one million were partially damaged. Longer-term financial losses have yet to be tallied from radioactive leakage, evacuation of hundreds of thousands of people, and ongoing attempts at cleanup from major explosions and meltdowns at four of the nuclear reactors at Fukushima Daiichi (One). The tsunami knocked out power from backup generators for the reactors. Damage from the shaking during the earthquake prevented outside power from being restored before three reactors overheated and melted down.

One minute before the earthquake was felt in Tokyo, an early-warning system sent out messages of impending strong shaking to millions of people and institutions in Japan, likely saving many lives. The concept behind this system and a few other similar systems in other parts of the world is to record and analyze the P and S waves close to a strong earthquake and to relay immediate warnings at the speed of light to more distant places before strong shaking reaches them. Warning times typically were ten to sixty seconds.

Although the rupture extended (figure 10.6) along the boundary between the Pacific and North American plates for about 300 miles (500 kilometers)

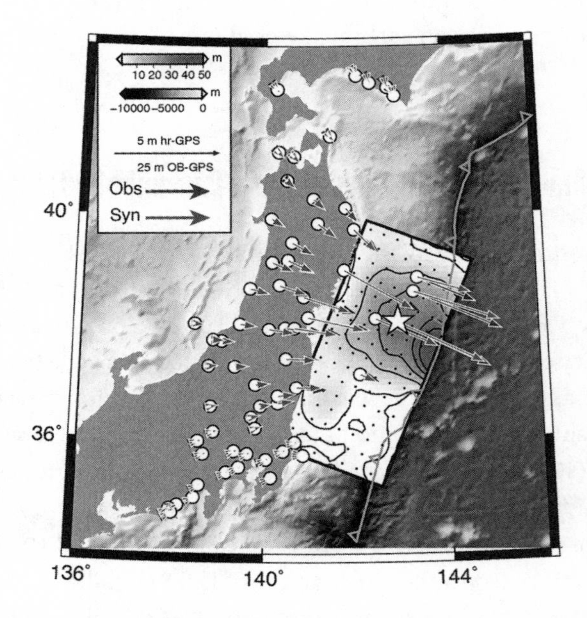

FIGURE 10.6 Fault displacements (horizontal slip) during the Tohoku mainshock of March 11, 2011. The large open star denotes its epicenter. Darker arrows denote GPS observations of slip; lighter arrows indicate calculated slip. The solid black box encloses the approximate rupture area of the mainshock. Darker shades indicate greater amounts of slip. Black indicates the deepest part of the Japan Trench, the easternmost extent of slip. See plate 17.

Source: Yue and Lay 2013.

parallel to the coastline, most of the slip was more concentrated along that boundary. It extended about 140 miles (225 kilometers) down the west-plunging plate boundary from its outcrop at the deepest part of the Japan Trench (dark shading in figure 10.6) to depths of about 40 miles (60 kilometers). The average fault displacement over that entire area was about 33 feet (10 meters). Horizontal slip close to the trench was an extraordinary 200 to 260 feet (60 to 80 meters). A shock of magnitude 7.9 about thirty minutes later, which undoubtedly was triggered by the mainshock, extended the rupture zone to the southwest. It, not the giant mainshock, caused the greatest damage to industries near Tokyo.

The mainshock was recorded by an exceptionally great number of seismograph stations and geodetic observation points both on- and offshore. Recent advances in computer power and modeling permitted many new things to be done with the geodetic and seismic data gathered in 2011. More than 2,000 technical papers have been published on this event, its aftershocks, and the phenomena preceding it. Those papers advance our understanding of earthquakes, the subduction process, and long-term prediction.

Figure 10.6 shows many of the GPS stations on land that measured horizontal displacements in 2011. This earthquake was unique in that Japanese scientists had installed five geodetic stations on the ocean bottom (referred to as OB-GPS) starting in 2002. One of those stations was located just above the hypocenter, the point where rupture initiated in 2011. Seismic data, offshore and onshore geodetic data, and tsunami records from buoys farther east in the Pacific provided an unprecedented opportunity to accurately map slip far offshore near the deepest part of the Japan Trench, where it was very large.

Prior to the event in 2011, it was difficult to estimate slip close to the deepest part of the Japan Trench using stations on land alone. That part of the trench is located farther from the coastline than it is for several other subduction zones, making estimates of slip far offshore in previous Japanese events very uncertain. Japan is now installing more geodetic and seismic stations offshore along its active plate boundaries, especially along the Nankai subduction zone.

Nearly all of the slip along the plate boundary during the mainshock of 2011 occurred within the black box in figure 10.6. Displacements varied considerably in that area, being largest at offshore stations within the shaded areas in the box between latitudes 37°N and 39°N. Fortunately, those offshore stations were installed before the earthquake in 2011, in part because Japanese scientists were expecting another shock of about magnitude 7.0 off Miyagi Prefecture, which, in fact, occurred two days before the mainshock of 2011. That large foreshock was followed in two days by additional moderate-size earthquakes whose epicenters approached that of the coming giant shock. Many geophysicists regard that forerunning sequence as a precursor to the giant earthquake. Until the mainshock occurred, however, those events were not identified as foreshocks.

Vertical displacements in the mainshock (not shown in figure 10.6) were considerably smaller than horizontal displacements since the rupture zone plunged (dipped) at a shallow angle to the west beneath the island of Honshu. Thus, the horizontal component of the slip along the plate boundary was much larger than the vertical component. Vertical motion of the seafloor during the shock was upward far offshore and down near of the coastline, as in figure 10.4. Subsidence near the coastline likely will persist for many decades or longer as stresses slowly build up to the next great or giant earthquake. A tsunami was generated by the vertical displacements of the seafloor. As the tsunami propagated closer to shore and encountered shallower water, its height and damage potential increased, causing most of the fatalities in 2011.

Northern Honshu is located on the North American plate, abbreviated NAM in figure 10.2. (Yes, the North American plate extends all of the way to northern Japan, although some researchers divide those parts of it west of the Aleutian Islands into what are called the Amur and Okhotsk microplates.)

In the years before the giant earthquake, points at the surface of the North American plate both offshore and on land (figure 10.7) moved northwesterly, exactly *opposite* to the southeasterly displacements during the mainshock of 2011 (figure 10.6). Those northwesterly displacements in the decades prior to the mainshock indicate that the North American plate was being compressed slowly as two nearby points on it were converging. Thus, the boundary between the North American and Pacific plates at depths where slip occurred in 2011 was locked at large strong patches called *asperities*. If the plate interface had instead been moving slowly at the long-term rate of plate motion prior to the giant earthquake, points on the North American plate near the subduction zone would not have moved closer together (as they did). Those compressive stresses were released

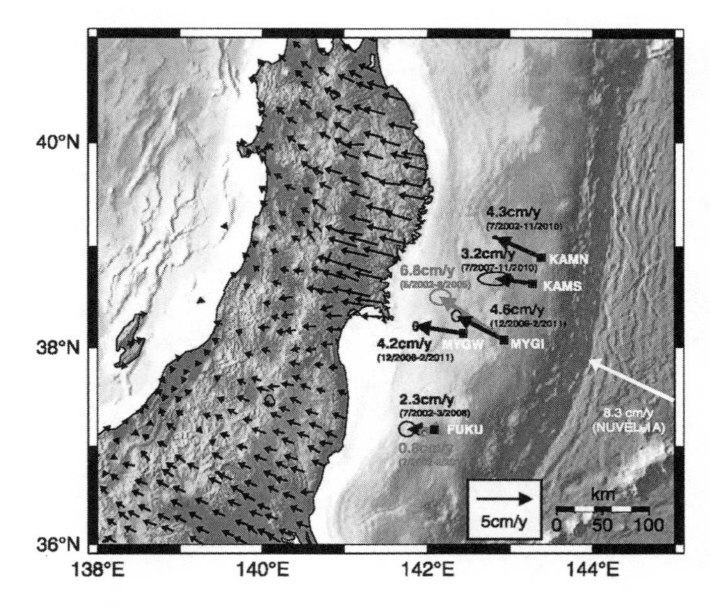

FIGURE 10.7 Horizontal displacements per year on land and at five stations on the seafloor before the 2011 earthquake. The white arrow at the lower right indicates the long-term rate of motion of the Pacific plate with respect to Honshu, 3.3 inches per year (8.3 centimeters per year). Note that point MYGI moved northwesterly at 82 percent of the plate rate between 2002 and 2005. See plate 18.

Source: After Sato et al. 2013.

suddenly during the giant shock. These patterns are like those reported in chapter 7 for the Chilean giant earthquake of 2010.

GPS measurements began at three offshore sites in 2002 (plate 18). The north*westerly* movement of point MYGI on the seafloor increased more rapidly from 2002 to 2005 (blue arrow) than in the several years prior to the last measurement (red arrow), a month before the giant earthquake. Point FUKU exhibited a similar behavior. Hence, the North American plate was not being compressed as fast between 2005 and 2011 as it was from 2002 to 2005.

In 2014, using GPS data from land stations, the graduate student Andreas Mavrommatis of Stanford University and two colleagues obtained similar results for a longer period from 1996 to just prior to the giant shock (figure 10.8) (Mavrommatis, Segall, and Johnson 2014). In 2015, Yusuke Yokota and Kazuki Koketsu reanalyzed that major long-term transient and found

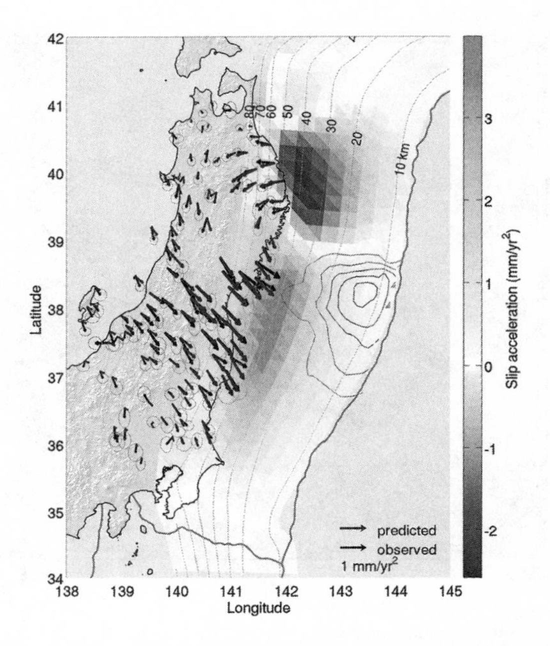

FIGURE 10.8 Slip on the plate boundary as deduced from GPS stations on land from 1996 until just before the giant earthquake of 2011. Slip accelerated in lighter areas and decelerated in darker areas. See plate 19.

Source: Mavrommatis, Segall, and Johnson 2014.

that it began about nine years before the giant 2011 event and was very large, magnitude 7.7.

Why did the northwesterly compression of the upper North American plate both offshore and on land decrease with time in the nine years before the shock in 2011? A good explanation is that slow slip (creep) increased with time along large parts of the plate boundary at depth.

Mavrommatis and his colleagues found in 2015 that the rate of repeating small earthquakes increased in the coming rupture zone of the shock in 2011 (Mavrommatis et al. 2015). Each small shock in a sequence occurred at nearly the same place. Their increased rates are consistent with the inference that slow seismic slip increased along a substantial part of the down-going plate boundary in the years before 2011. These observations of possible precursory increases in the rate of slip along the plate boundary are very exciting for long-term earthquake prediction.

The giant Japanese earthquake of 2011 was unusual in that slip took place on the plate boundary all the way out to the deepest part of the Japan Trench. Han Yue and Thorne Lay (2013) found that slip along the part of the rupture zone close to the trench generated weak short-period seismic waves and large longer-period waves (darker-gray area in figure 10.9). Slip farther west closer to land at greater depths in the lighter-gray area generated stronger short-period seismic waves. Lay and his colleagues found that slip during the giant earthquake off Sumatra in 2004, which extended all the way out to the deepest part of its trench, also was characterized by a similar distribution of poor and abundant short-period seismic waves. The presence of large long-period but weak short-period waves indicates slip occurred more uniformly and perhaps more slowly close to those trenches and contributed to the generation of particularly huge tsunami waves.

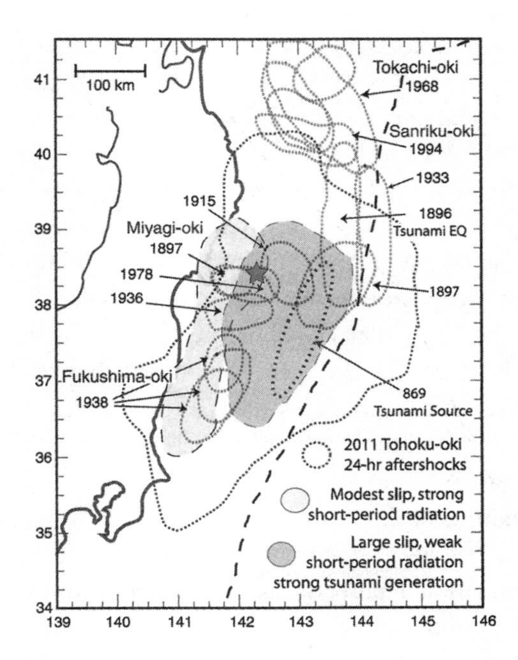

FIGURE 10.9 Rupture areas of 2011 Tohoku–Oki earthquake that generated strong, short-period seismic waves (*lighter shade*) and waves with large slip but weak short-period radiation (*darker shade*) closer to the Japan Trench. The thick dashed black line indicates the deepest part of the trench. The star denotes the epicenter. See plate 20.

Source: Koper et al. 2011.

In contrast, in many great, but not giant, shocks at subduction zones, rupture occurs along those parts of plate boundaries closer to coasts, as in the lighter-gray area of figure 10.9, and does not extend as far seaward as the deepest parts of trenches. For example, rupture in the great Iquique, Chile, earthquake of 2014 did not extend as far west as the Peru–Chile trench (figure 7.5). The tsunami it generated was not as large as those for the giant earthquakes of 2004 and 2011.

Koichiro Obana of the Japan Agency for Marine-Earth Science and Technology and thirteen colleagues operated ocean-bottom seismometers close to the Japan Trench soon after the shock in 2011. Most of the earthquake mechanisms they found involved either normal or strike-slip faulting in the overriding and incoming plates, but not along the plate boundary itself. Shocks prior to the earthquake, however, involved thrust and strike-slip mechanisms. Hence, stresses closer to the trench likely were dropped to zero along that part of the plate interface during the giant shock.

In 2014, my Lamont colleague Christopher Scholz argued that the earthquake in 2011 began as a fairly typical great subduction-zone event, initiating at a depth of about 12 miles (20 kilometers) and some 60 miles (100 kilometers) landward (westward) from the deepest part of the Japan Trench. Rupture then propagated closer to shore off Miyagi and Fukushima prefectures (the lighter-gray area of figure 10.9). About fifty seconds later, a huge surge in slip began about 45 miles (60 kilometers) from the trench at shallow depth and propagated all the way to the deepest part of the trench. Scholz argued that the paucity of short-period seismic waves suggests a huge, smooth, and uninhibited surge in slip of more than 165 feet (50 meters) close to the trench, as designated by the darker-gray area of figure 10.9. The preceding slip in the lighter-gray region likely triggered the huge surge in the darker-gray area.

The huge amounts of slip in the darker-gray area of figure 10.9 indicate that it does not break very often in conjunction with a great shock farther west. The accumulation of potential slip of 165 feet (50 meters) by slow plate motion would take about 600 years. In fact, the region shown as the darker-gray area appears to rupture about every 600 years, as deduced from historic giant tsunamis. Therefore, the giant shock of 2011 consisted of two quite different types of slip.

The approximate rupture zones of the great Japanese shocks of 1896 and 869. are superposed on the distribution of slip during the 2011 earthquake in figure 10.9. In 1972, Hiroo Kanamori of Cal Tech called the 1896 event a "tsunami earthquake" because it generated a very damaging tsunami but caused only relatively small seismic shaking along the coast of northeastern Japan. He placed the earthquake close to the Japan Trench. The northeastern part of the 2011 rupture zone overlaps the 1896 rupture zone.

The earthquake of 2011 provides some lessons for great and giant shocks along the Cascadia subduction zone off the coasts of northern California, Washington, Oregon, and adjacent British Columbia. The Cascadia zone ruptured in a giant earthquake in 1700, which produced a huge tsunami that affected Japan. The event in 1700 appears to be much like the Japanese shock of 2011 in its long rupture zone parallel to the coast, magnitude near 9.0, and its huge tsunami. Paleoseismic work along the coasts of Cascadia indicates similar amounts of vertical motion in 1700. Paleoseismic investigations identified previous great and giant shocks along Cascadia, some as large as that of 1700, but others that broke somewhat shorter segments. Thus, the amounts of damage in the earthquake of 2011 and the size of its tsunami need to be taken into account in estimating maximum hazards for Cascadia. Cascadia shocks somewhat smaller than that of 1700, as detected by paleoseismic work, could affect coastlines as long as that of either Oregon or Washington.

Damage to the Kashiwazaki-Kariwa Nuclear-Power Plants from a Large Earthquake Along the West Coast of Japan in 2007

Prior to the reactor accident at Fukushima One, Japan had fifty-four operating nuclear-power reactors at sixteen sites, providing about 30 percent of the country's electricity. Japan is the world's third-largest producer of electricity from nuclear energy. Tokyo Electric Power Company (TEPCO), a huge enterprise, owns and operates seventeen reactors: six at Fukushima One, four at Fukushima Daini (Two), and seven on the west coast of Honshu at Kashiwazaki-Kariwa. Because Japan has little coal and petroleum that can be used to generate electric power, it turned to nuclear power for its electrical needs even in the face of risks from earthquakes and tsunamis. The nuclear industry has resisted the installation of large wind farms for electrical generation in Japan.

The seven nuclear-power reactors owned by TEPCO along the west coast of the main Japanese island of Honshu were strongly shaken by an earthquake of magnitude 6.7 in 2007. Those reactors, which are located between the towns of Kashiwazaki and Kariwa, produced more electrical power than any other site in the world.

This shock involved thrust faulting along the boundary between the Eurasian and North American plates, where the crust and mantle beneath the Japan Sea to the west is being underthrust (subducted) easterly under the west coast of Honshu. The rate of subduction, about 0.4 inches (one centimeter) per year is much less than along the east coast of Honshu, the site of the 2011 earthquake. Nevertheless, a number of large to great shocks, including the damaging Niigata earthquake of magnitude 7.5 in 1964 to the northeast of the reactors, occurred along that plate boundary. It was fortunate that the shock of 2007 was not as large as that of 1964.

Initial studies of the Kashiwazaki-Kariwa site prior to construction examined faults at the surface on land but seem to have ignored the plate boundary that plunges under the reactor site and outcrops on the seafloor just to the west. How the existence of the plate boundary and the occurrence of the Niigata earthquake in 1964 were ignored in designing those reactors is puzzling. Seismologists in Japan were well acquainted with the Niigata shock. Construction of the first reactor started in 1980. *Wikipedia* states that the reactors were shut down one by one in 2002 and 2003 after the discovery of the deliberate falsification of data.

All seven of the reactors at Kashiwazaki-Kariwa experienced shaking in 2007 that was 1.4 to 2.5 times larger than they were designed to withstand. Small amounts of radioactive water were spilled, but no major radioactive releases affected the public. Because fire engines were not kept on site, a fire in the complex took some time to extinguish. The administration building and a tall chimney at the site were badly damaged during the earthquake, whose epicenter was only 12 miles (19 kilometers) away.

The reactors at Kashiwazaki-Kariwa were out of service for more than twenty-one months, with a loss of electrical generation of roughly U.S.$3 billion and a shortage of electricity in Tokyo. Those units, like the ones at Fukushima, are boiling-water reactors similar to those made by General Electric in the United States. The reactors at Kashiwazaki-Kariwa were shut down again in 2011 following the Fukushima disaster; some were to be restarted at the time this book was in publication.

Major Accident at the Fukushima One Reactors in 2011

Prior to 2011, thinking and policy decisions about reactor design in Japan concentrated on earthquakes and tsunamis that occurred only within the past 120 years. The seismic magnitudes of those shocks were not larger than 8.0 to 8.5 for northern Honshu and considerably smaller for the rupture zone where the earthquake occurred in 2011. The sizes of tsunamis and earthquakes that those events generated were used as standards for constructing nuclear-power reactors and seawalls to prevent tsunami damage.

Nevertheless, several earth scientists in Japan knew about the great Jogan earthquake of 869 for at least a decade before 2011. Its damaging tsunami was comparable to that of 2011 and in the same general area. One community in northern Honshu used engraved stones to mark the farthest inland reach of the tsunami in 869. They remembered those stones as an indication of how far uphill they should flee in a future giant shock, which they did in 2011.

A tsunami generated by another great earthquake in the seventeenth century farther northeast along the plate boundary flooded farther inland

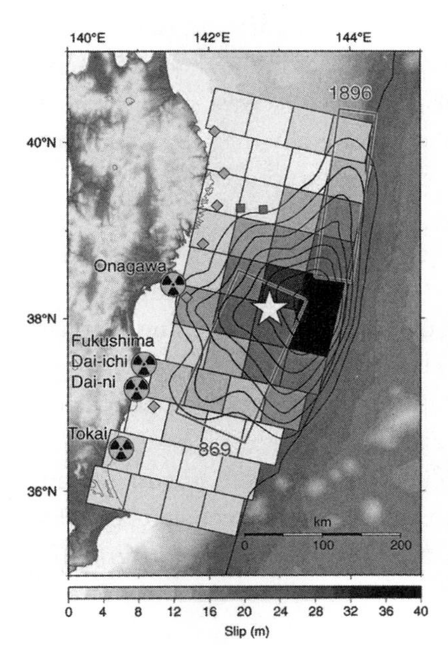

FIGURE 10.10 Cumulative slip in the giant earthquake of 2011; the large star indicates the epicenter. Radiation symbols mark locations of four nuclear-power plants. The outer line is the deepest part of Japan Trench. See plate 21.

Source: Satake et al. 2013.

along the northern Japanese island of Hokkaido than tsunamis generated by shocks of the past 120 years. As in 2011, this seventeenth-century shock may have involved slip that also extended out to the deepest part of the trench.

The shock of magnitude 9.0 in 2011 occurred before a consensus developed in Japan about including the giant Jogan shock of 869 in hazards assessments. The damage done by the earthquake in 2011 and its magnitude came as a surprise to most scientists and to the populace of Japan. The tsunami was larger than anticipated, overtopping tsunami walls in several places along the northeastern coast of Honshu. For reactors, this shock was a "beyond-design-basis" earthquake.

The occurrence of the giant earthquake of 2004 off Sumatra, Indonesia, and its devastating tsunami led earth scientists in Japan to reassess the tsunami generated by the Jogan earthquake off Tohoku. Calculations in 2008 and 2010 using tsunami data for the Sumatran shock indicated that the rupture zone of the Jogan earthquake had a width and a length of about 62-by-125 miles (100-by-200 kilometers), still smaller than the dimensions of the rupture in 2011. Nevertheless, those calculations were a step in the right direction in terms of disaster assessment. Understandably, little was known and is still unknown about slip near the Japan Trench in 869.

Planning in Japan before 2011 did not consider an earthquake that could rupture both the plate boundary near the trench, as in the shock of magnitude 8.3 in 1896, and the plate boundary farther to its west, as in the magnitude 8.2 event of 1968. Values closer to 9.0 are obtained if all segments from the coast to the trench are assumed to have ruptured in 869, as they did in 2011.

In 2011, the earthquake and tsunami struck four nuclear-power stations along the coast of northeastern Honshu (figure 10.10). The tsunami height at the Fukushima Daiichi (One) complex of six reactors, where huge damage occurred, was 43 feet (13 meters), considerably higher than the design height of 20 feet (6.1 meters). The tsunami reached the Fukushima One reactors forty-one minutes after the initiation of the earthquake.

In 2013, Kenji Satake and colleagues indicated that the tsunami designs for Fukushima One and Two were based on a series of earthquakes offshore of Fukushima and Ibaraki prefectures in 1938. Those events of magnitudes 7.4 to 7.5 were the largest shocks in that area during past 120 years. Design magnitudes of 7.4 to 7.5 seem much too small given the occurrence of shocks of magnitude 8.0 and larger to the north along the plate boundary. In fact, in 2008 the Headquarters for Earthquake Research Promotion in Japan stated that a shock of magnitude 8.2 could occur anywhere along the Japan Trench.

For more information on the Fukushima disaster, I recommend the two reports by the U.S. National Academies of Sciences, Engineering, and Medicine, *Lessons Learned from the Fukushima Nuclear Accident for Improving Safety of U.S. Nuclear Plants* (2014, 2016) and the book *Fukushima: The Story of a Nuclear Disaster* (2014) by David Lochbaum and his colleagues of the Union of Concerned Scientists. The lessons for the United States from Fukushima are discussed further in chapter 13.

The National Academies report published in 2014 indicates that when the earthquake occurred in 2011, two reactors at the Onagawa nuclear complex (figure 10.10), the closest reactors to the rupture zone, were at full power, and a third was warming up. All three reactors stopped generating electricity (scrammed) automatically. The earthquake tripped four of five off-site electrical power lines to the site. Two of five on-site power generators shut down after the tsunami flooded their cooling-water pumps. Fortunately, however, the other three generators supplied power to the site until off-site power was restored the next day. If that had not happened so quickly, another disaster like Fukushima One might well have occurred.

The maximum tsunami height at the Onagawa complex was 43 feet (13 meters), which was lower than the elevation of the main plant. Hence, most of the facility did not experience flooding. The earthquake and tsunami caused some damage to the plant but not to its structural integrity; no meltdowns occurred. Fortunately, the first reactor at Onagawa was designed in 1974, eight years after the first reactor at Fukushima One. With time, TEPCO had increased the design heights of plants to withstand tsunamis for various reactors in northeastern Honshu after the first construction permits were issued for units at Fukushima One in 1966.

The National Academies report of 2014 states that four of the six nuclear reactors at Fukushima One were situated at an elevation of 33 feet (10 meters). The elevations of those four reactors were well below the tsunami height of 43 feet (13 meters) in 2011. The elevation of a breakwater designed for tsunami protection was only 13 feet (4 meters), and the elevation of the emergency diesel generators for units 1 to 5 was only 6.6 feet (2 meters).

One reactor at the Fukushima Two complex was operating at the time of the earthquake in 2011. The shock cut off outside AC power, but emergency generators supplied backup power until off-site power was restored two days later. The maximum tsunami height at Fukushima Two was below the elevation of the main part of the plant, so Fukushima Two didn't experience flooding. The operating reactor reached cold shutdown two days after off-site power was restored. Again, if off-site power had not been restored that quickly, the operating reactor may have melted down, as happened at Fukushima One.

Units 1 to 3 at Fukushima One were generating electricity at their licensed designs at the time of the earthquake; unit 4 was shut down for replacement of its reactor core; and units 5 and 6 were out of commission for inspection, with their reactors being actively cooled. All six of the reactors successfully "scrammed"—that is, they stopped producing electricity when P waves from the earthquake shook the site. Nevertheless, a scram, which involves inserting control rods into a reactor, is only the first step in arriving at a cold safe shutdown and preventing the meltdown of a reactor's nuclear fuel. Electrical power is needed to cool the fuel rods after a reactor stops producing electricity by nuclear reactions.

Within two minutes, shaking from the earthquake cut off-site AC electrical power needed to cool reactors at Fukushima One, well before the arrival of the tsunami about forty-one minutes later. Hence, the damage to Fukushima One cannot be attributed solely to the tsunami, as some have claimed. Emergency generators started to supply backup power for cooling until they were knocked out by the tsunami. The main reason the tsunami resulted in so much damage at Fukushima One was that the plant and its backup electrical generators were situated at the lowest elevation of the four reactor sites shown in figure 10.10. Loss of power resulted not only in meltdowns at three reactors but also in the operators' inability to read critical instruments, access safety systems, and control the reactors.

In the seventy-two hours following loss of outside power, the reactors at units 1, 2, and 3 melted down, releasing highly radioactive materials. Hydrogen explosions in the reactor buildings of units 1, 3, and 4 caused severe structural damage and the release of high levels of radioactive materials. Hydrogen leaked from unit 3 into unit 4 through a connecting pipe. Hydrogen, a flammable and explosive gas, is produced when the zirconium cladding surrounding nuclear fuel is heated to high temperatures. An emergency diesel generator at unit 6 was undamaged. It supplied power to units 5 and 6, both of which reached cold shutdown nine days after the earthquake and tsunami.

Outside power was not restored to Fukushima One until March 18, well after meltdowns of and damage to four reactors. TEPCO neither expected nor planned for a prolonged lack of both off-site and on-site electrical power. The loss of electrical power also shut down the plant's radiation-monitoring system.

In 2014, the National Academies report made five other recommendations to help prevent hydrogen explosions and meltdowns and to bring reactors to cold shutdowns during extreme "beyond-design-basis" events. The last reactor at Fukushima One did not reach cold shutdown until about nine months after the earthquake.

The academies report states that TEPCO, the owner of Fukushima One, seemed to ignore the urgency to make any changes in 2008 when wave heights from the tsunami of 869 were modeled as 33.5 feet (10.2 meters), much larger than the design tsunami of 20 feet (6.1 meters) used for units 1 to 4. According to the report, TEPCO merely called for more studies, an action that I have found a familiar refrain in the United States for existing, older nuclear-power plants.

The academies report also states, "The overarching lesson learned from the Fukushima Daiichi accident is that nuclear plant licensees and their regulators must actively seek out and act on new information about hazards that have the potential to affect the safety of nuclear plants." It goes on to say, "Nuclear plants usually operate for many decades. Scientific understanding of hazards, especially hazards arising from natural external events, can advance substantially during such extended periods."

When new information indicated a greater earthquake risk for reactors in the United States designed more than 40 years ago, the overwhelming view within the U.S. Nuclear Regulatory Commission (NRC) was to keep older reactors operating and to claim they are safe. Although it was not realistic to raise the heights of reactors at Fukushima One after they were built, higher tsunami walls could nevertheless have been constructed and on-site emergency generators could have been moved to higher ground to prevent flooding during a large tsunami. Those measures were not taken.

These and other lessons from Fukushima One have largely been ignored thus far by operators of nuclear-power reactors in the United States and by the NRC. The authors of the National Academies reports as well as David Lochbaum and his colleagues are much more knowledgeable, wise, and cautious in their assessments than the NRC documents I have seen thus far.

The 2014 and 2016 National Academies reports discuss extensively the many measures that were taken to bring the damaged reactors at Fukushima under control after the tsunami hit. The first report states, "Indeed, the events at the Fukushima Daiichi plant demonstrate the extraordinary difficulty of executing a successful response to accidents involving multiple reactor units under the difficult conditions that existed at the site. . . . The extensive damage and [radioactive] contamination were totally unexpected by the operators at the Fukushima Daiichi plant." In contrast to the presence of many reactors at single locations in Japan, only one site in the United States consists of three or more reactors. The other U.S. sites contain either only one or two operating reactors.

The National Academies report of 2014 did find "that the Fukushima accident was not a technical surprise and was in fact anticipated by previous severe accident analyses." It observes that "the possibility of a reactor-core-damaging event at a nuclear plant in Japan was considered implausible. Consequently, planning for such an event was not treated seriously, leaving Japan unprepared for the scope and extent of the required emergency response."

The Fukushima accident highlights concerns about the vulnerability of spent-fuel storage pools at reactors. Large amounts of spent fuel, which is highly radioactive, were stored at the reactors at Fukushima One at the time of the earthquake in 2011. They were located in the main-reactor buildings but outside of primary containment structures. Storage of spent fuel high up in individual units ("in the attics") is a serious design flaw for reactors manufactured by General Electric and for similar units built in Japan. For some time after the earthquake, Japanese workers were not able to pump water to those heights at Fukushima One.

In late 2014, the NRC ruled against a petition to close reactors of the General Electric type in the United States. It also has permitted greater amounts of highly radioactive spent fuel to be stored in pools at U.S. reactors, which the Japanese have not.

The Fukushima accident resulted in the most extensive release of radioactive materials into the environment since the Chernobyl nuclear-power accident of 1986 in the Soviet Union. High levels of radiation either slowed or halted many efforts to lessen damage at Fukushima One. The most potent releases from the Fukushima and Chernobyl disasters involved the radioactive isotopes iodine-131 and cesium-137. The first has a half-life of only eight days but can be very destructive to the thyroid gland, especially in children. Cesium-137, whose gamma radiation is powerful and damaging to humans and other animals, has a half-life of 30 years. It is a common decay product of the fissioning of uranium.

Cesium reacts with water to form cesium hydroxide, which is soluble in water and difficult to remove from soils, forests, and vegetation. The biological behavior of cesium is similar to that of the elements potassium and sodium. Cesium soon became the principal source of radiation and the greatest risk to health for time scales of years to decades following the Chernobyl accident, as it will be for the Fukushima accident.

According to Lochbaum and colleagues (2014), "about 350,000 people were resettled; an area of radius of about 19 miles (30 km) around the [Chernobyl] plant remains an 'exclusion zone.'" An epidemic of thyroid cancer in children in Ukraine and parts of Belarus and Russia followed in the years after the Chernobyl disaster from the ingestion of milk and other foods containing iodine-131.

Prevailing winds at the time of the Fukushima accident initially blew about 80 percent of the released radiation into the Pacific Ocean. Winds subsequently shifted and carried radioactive materials over land along a band stretching about 30 miles (50 kilometers) to the northwest. Some of the hotter radioactive spots on land occurred in that band more than 12 miles (20 kilometers) from the reactors. In contrast, most radioactive releases from Chernobyl were deposited over land.

The Japanese government is under great pressure to clean up radioactive cesium in the broad area of Fukushima and neighboring prefectures so that people who were evacuated can return to their homes and livelihoods. The National Academies report of 2014 states that more than 80,000 people continued to live in temporary locations three years after the disaster.

The environmental and economic consequences of the nuclear accident in 2011 continue to be severe. The National Academies report of 2014 states that about 5,000 square miles (13,000 square kilometers) of land, about the size of the state of Connecticut, are contaminated at dose levels that exceed those of the long-term cleanup goal. Disposal of radioactive waste is proving to be difficult. A determination of how much land will be off limits indefinitely still has to be made as of late 2018.

11

EARTHQUAKES IN THE EASTERN AND CENTRAL UNITED STATES

As a young assistant professor, I received funding in 1969 from the Sloan Foundation to install fifteen seismograph stations in New York State, northern New Jersey, and Vermont. At that time, very few modern stations were in operation in the eastern United States, and little was known about the distribution of earthquakes and seismic risk.

The large region between the west side of the Rocky Mountains and the western Atlantic Ocean past the east coast of the United States as far as the Mid-Atlantic Ridge is an intraplate area—that is, an area in the middle of the North American plate (plate 1). Although seismic activity in this intraplate area is lower than the activity along most plate boundaries, many earthquakes have occurred here close to centers of population and major structures. In addition, shaking occurs over much greater distances in the eastern and central United States than in California and Nevada because seismic waves propagate more efficiently there. Crustal stresses are generally high in this area for reasons that we poorly understand. It is important to note that most of the nuclear-power plants in the United States and Canada are situated east of the Rockies in this intraplate region.

Earthquakes in New York, New England, New Jersey, and Pennsylvania

As part of his Ph.D. work at Lamont, Brian Isacks operated very sensitive seismographs in the early 1960s within a deep mine at Ogdensburg, New

Jersey. Seismic noise was low enough that he was able to record high-frequency seismic waves from earthquakes at local and regional distances. Because the largest seismic waves at these distances are high-frequency waves, he detected many small earthquakes.

The stations we installed with funding from the Sloan Foundation also had excellent high-frequency capabilities. The Lamont network eventually grew to about fifty stations and is now overseen by Won-Young Kim. High-frequency seismic waves are now being recorded at stations around the world to better monitor the Comprehensive Nuclear Test Ban Treaty of 1996.

Figure 11.1 shows earthquakes detected by seismograph stations in New York State and adjacent areas. Northern New York, including the Adirondacks, is the most active part of the state. The Cornwall, Ontario, and Massena, New York, earthquake of magnitude 5.8 in 1944 caused considerable

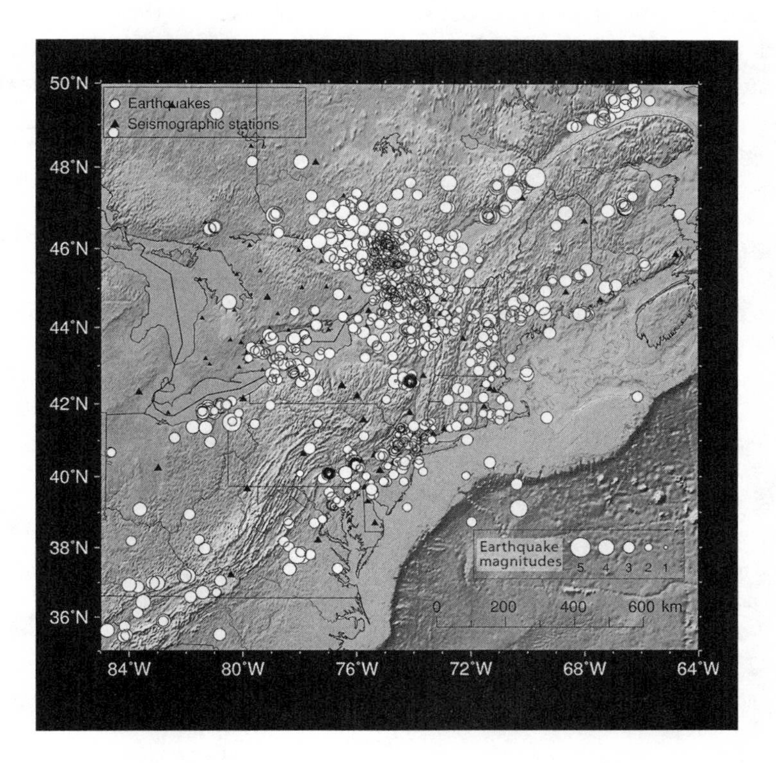

FIGURE 11.1 Earthquakes of magnitude 2.5 and greater in the mid-Atlantic states, New England, and adjacent parts of Canada from 1990 to 2003. See plate 22.

Source: Lamont-Doherty Earth Observatory unpublished report, 2003.

damage in those two cities and was felt as far away as Maryland and Michigan. That event is part of a larger seismic zone that extends northwesterly into Canada, where several large shocks have occurred in the vicinity of Montreal and Ottawa. In 1935, the Timiskaming earthquake of magnitude 6.4 farther to the northwest was felt as far south as Washington, D.C., west to Wisconsin, and east into eastern Maine.

The second most active region depicted in figure 11.1 extends from southwestern Connecticut across the New York City metropolitan area to south-central New Jersey and southeastern Pennsylvania. It is the most populated region in the United States. Western New York State is the third most seismic region. It was the site of the Attica earthquake of 1929 and of a shock near the New York–Pennsylvania border in 1998, each of magnitude 5.2. The central part of New York State and most of Pennsylvania are characterized by lower earthquake activity.

Several of us at Lamont focused on earthquakes in the greater New York City–Philadelphia area; figure 11.2 shows events of magnitude 3.0 and greater in that area since 1700. The locations and magnitudes of nearly

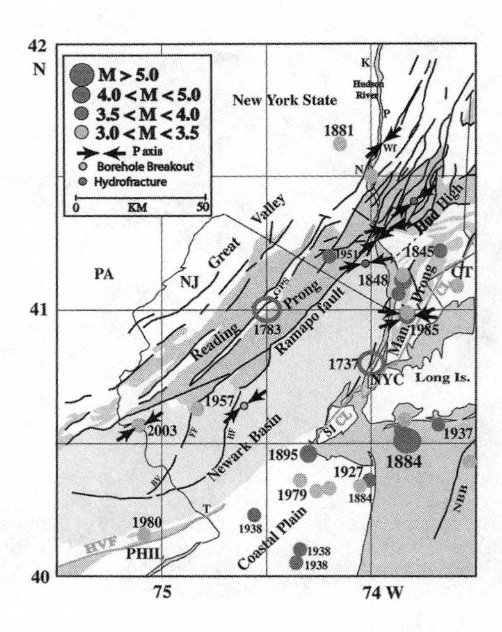

FIGURE 11.2 Earthquakes in the greater New York City–Philadelphia area, 1700 to 2004. See plate 23.

Source: Sykes et al. 2008.

all earthquakes here prior to 1950 are based on reports of shaking, mainly from old newspapers.

The largest known earthquake (magnitude 5.25) in the area occurred in 1884 just offshore of Brooklyn, New York. The exact locations of the next-largest shocks of magnitude 5.1 in 1737 and 1783 are poorly known, with an uncertainty of about 30 miles (50 kilometers). I made a search of electronic copies of old newspapers and surprisingly found more reports for the event of 1737 than for that of 1783.

On December 19, 1737, a strong earthquake was felt from New Hampshire to New Castle, Delaware. (This event was reported on December 7 according to the older Julian calendar that was in use at the time.) Three severe "shocks of an earthquake" were felt at New York City, but only one event was strongly felt in Philadelphia and New Castle. One newspaper reported that this shock threw down several chimneys and bricks from a wall and that bells tolled in church steeples in New York City.

The report on the earthquake in New York City in 1737 (see figure 11.3) states, "Here we had first one single Shock, and after an Intermission of a few Seconds a violent Tremor." The one single shock can be interpreted as the P wave that was followed by the S wave from the same earthquake after a few seconds. If so, the main event of 1737 must have occurred in or close to New York City.

On November 30, 1783, three earthquakes occurred at the end of the American Revolutionary War, a few days after the entry of American troops

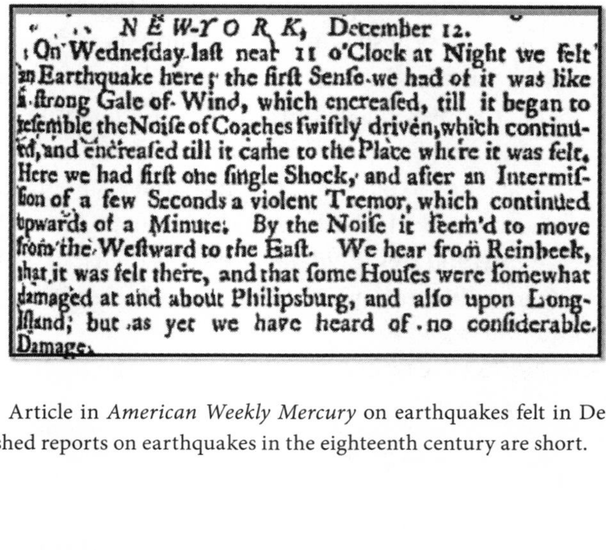

NEW-YORK, December 12.
On Wedneſday laſt near 11 o'Clock at Night we felt an Earthquake here; the firſt Senſe we had of it was like a ſtrong Gale of Wind, which encreaſed, till it began to reſemble the Noiſe of Coaches ſwiftly driven, which continued, and encreaſed till it came to the Place where it was felt. Here we had firſt one ſingle Shock, and after an Intermiſſion of a few Seconds a violent Tremor, which continued upwards of a Minute. By the Noiſe it ſeem'd to move from the Weſtward to the Eaſt. We hear from Reinbeek, that it was felt there, and that ſome Houſes were ſomewhat damaged at and about Philipsburg, and alſo upon Long-Iſland; but as yet we have heard of no conſiderable Damage.

FIGURE 11.3 Article in *American Weekly Mercury* on earthquakes felt in December 1737. Most published reports on earthquakes in the eighteenth century are short.

into New York City and the departure of the British. Many of the early newspapers that were owned by British sympathizers were no longer being published, as they were in 1737. Understandably, more attention was devoted in 1783 to celebrations than to the series of three shocks felt in one day in New York City. One shock was "severely felt" but did little damage. Reports mention one earthquake felt as far away as New Hampshire and southeastern Pennsylvania and two felt in Hartford and Philadelphia. Thus, the three earthquakes must have occurred closer to New York City.

Since the Lamont seismic network became operational in 1973, several notable, moderate-size earthquakes were well located using instrumental data. On Saturday, October 19, 1985, I was abruptly shaken awake in my home in Palisades, New York, at 6:25 a.m. by an earthquake of magnitude 4.1 centered just across the Hudson River beneath Ardsley, New York. The shaking caused nails in my old wood-frame house to move and to generate loud noises that lasted at least a minute. Approximately 30 million people felt the shock.

I knew my day would be different from what I had planned originally. Quickly showering and dressing, I went to the Lamont Observatory, a mile away, where at 7:00 a.m. several TV trucks were already waiting at the entrance to interview us. Fortunately, we had a plan in place. Some scientists analyzed the data from the local network; others fielded phone calls; and yet others dealt with the great media interest. I requested that Lamont security send the press to an auditorium in another building, where they would not bother those of us analyzing the data. We announced right away that we would hold a press conference at 9:00 a.m., which allowed the media to make their evening deadlines. By then, we had estimates of the location and magnitude and reports of shaking from around the area.

We had devised this plan because we had a bad experience two years earlier, in 1983, after an Adirondack earthquake of magnitude 5.1, which was widely felt. One member of the press had roamed the building and pulled out electrical cords to our recording instruments to power his own equipment. That earthquake had attracted about thirty TV stations from New York, New Jersey, and New England. We knew then that we needed to have a plan in place for our interactions with the media after future earthquakes.

Enough aftershocks were recorded in 1985 that we could tell that the Ardsley earthquake occurred at a depth of about 3 miles (5 kilometers) along the Dobbs Ferry fault trending west–northwest, which had moved during the closure of a previous ocean about 450 million years ago and again during the early opening of the Atlantic Ocean 200 million years ago. The mechanism for the 1985 earthquake involved strike-slip faulting but without any displacement at the earth's surface.

On September 11, 2001, our Palisades seismograph station recorded the impacts of the two hijacked airlines on the World Trade Center and the subsequent collapses of the center's two main towers. Palisades was the closest seismic station and the only one to record the two impacts. The signals from the impacts were smaller than the signals from the collapses. Many stations in the Mid-Atlantic and New England area recorded the largest collapse at magnitude 2.4. Several of us at Lamont, with Won-Young Kim as first author, published the seismic results in 2001 (Kim et al. 2001). In 2016, Kevin Krajick, a writer for the Earth Institute of Columbia University, described in detail the work of Lamont scientist Kim and others after this event.

More Frequent Earthquakes in Stronger Rocks

Our work at Lamont showed that historical as well as more recent well-recorded earthquakes near New York City occurred more frequently in the older, stronger rocks beneath what is called the Manhattan and Reading prongs (figure 11.2) and not beneath the younger weaker rocks of the Newark basin. We think that earthquakes also occur in similar older, strong rocks at a depth below the young weaker rocks of the coastal plain. Shocks may occur more frequently in strong, older rocks for two reasons: (1) higher stresses can build up in them, and (2) they likely contain more old faults of various orientations that can be reactivated by the current stress regime.

Dilatancy Theory of Earthquake Precursors: Blue Mountain Lake, New York

I attended the meeting of the International Union of Geodesy and Geophysics in Moscow in 1971. Months before the meeting, I wrote to Igor

Nersesov seeking to visit one of two Complex Seismological Expeditions he headed in Central Asia. The day of my departure for Moscow I received a letter stating that I was to be the guest of the Soviet Academy of Science in Garm, Tajikistan, a major center for investigations of earthquakes and prediction. The Soviet Union had commenced field recording and analysis of earthquakes there following a very deadly shock near Garm in 1949.

In Garm, I learned about a technique they were using for local (nearby) earthquakes to measure the travel times of S waves divided by those for P waves. They claimed that this ratio changed significantly before several large earthquakes and that it may be useful for intermediate-term prediction. When I returned to the United States, I spoke about their work at a joint meeting on earthquakes in Aspen, Colorado, that included scientists from seismology and rock mechanics. My talk elicited much excitement.

A sequence of small earthquakes called a *swarm* occurred in the central Adirondacks at Blue Mountain Lake, New York, in 1971 and then again in 1973. It was centered near several of the seismograph stations that Lamont had installed around 1970. Lamont graduate student Yash Aggarwal worked on those sequences and found similar changes in the ratio the Soviets had reported.

Christopher Scholz, Aggarwal, and I next examined reported changes in that velocity ratio, sea level, and the radioactive gas radon before several other earthquakes in various parts of the world. We proposed that those changes took place in response to a phenomenon called *dilatancy*, which occurs when rocks are close to failure and cracks tend to close. We published our hypothesis in *Science* in 1973 (Scholz, Sykes, and Aggarwal 1973).

We later found that the Soviets' results were misleading because the small earthquakes before large shocks that they used had occurred at different times and places in rocks of quite different ages and seismic velocities. The changes they reported now appear to be related to their sampling of different rock types and not to changes in the velocity ratio. Several people in California looked at repeated industrial explosions whose seismic waves passed near moderate-size earthquakes. Most seismologists took their negative results as a failure of our dilatancy hypothesis. I became convinced that measuring those changes was a very difficult "row to hoe" because seismic sources were likely needed very close to moderate and large earthquakes, perhaps within major fault zones.

Our results from Blue Mountain Lake were unusual in that rocks there, unlike those in California, are strong very close to the earth's surface and may have undergone changes in dilatancy. Although our dilatancy hypothesis for earthquake prediction is unproven, it likely is not correct for the very large areas and velocity changes that we proposed in 1973. Stresses may not be high enough near plate boundaries to generate dilatancy, but they may be high enough in intraplate regions such as Blue Mountain Lake.

Other Notable Earthquakes in Eastern and Central North America

Earthquakes have occurred in the broad intraplate region of the United States and adjacent areas of Canada since 1700 (figures 11.1 and 11.4).

The earthquake history of the southeastern United States is dominated by the large Charleston, South Carolina, shock of about magnitude 7.0 in 1886, which occurred before the technology to make seismic recordings was developed. This shock caused considerable damage to many poorly constructed buildings in Charleston and killed sixty people. Damage was particularly high to structures built on the extensive soft sediments beneath the city. Columbia, South Carolina, and Savannah, Georgia, each about 90 miles (145 kilometers) from Charleston, also were strongly shaken. The shock, in fact, was felt over a huge area of eastern North America, including New York City and Boston, as well as in Bermuda and Cuba. In comparison, an earthquake of the same magnitude occurring west of the Rockies would have been felt within a considerably smaller area.

What fault or faults were active in this event in 1886 is still under investigation. Small earthquakes recorded by a local seismic network in South Carolina have occurred mostly in harder crystalline basement rocks beneath the younger sedimentary rocks of the coastal plain, much as what we think happens beneath the coastal plains of New Jersey and New York (figure 11.2).

An earthquake of similar size occurred in 1929 beneath the Grand Banks off the coasts of Newfoundland and Nova Scotia. It triggered a tsunami that affected those coasts with a loss of life at Placentia Bay, Newfoundland,

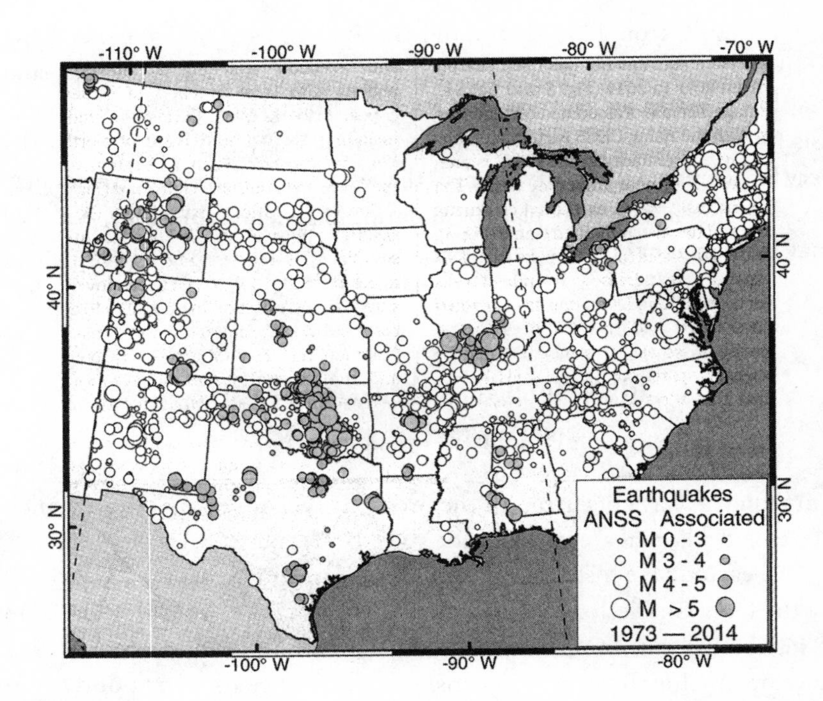

FIGURE 11.4 Earthquakes within the North American plate 1973–2014. Solid symbols indicate earthquakes triggered by fluid injection. See plate 24.

Source: Weingarten et al. 2015.

and an underwater landslide of sediments called a *turbidity current* that damaged twelve submarine cables in the Atlantic.

It is not known if the areas of the Charleston and Grand Banks earthquakes are more prone to major shocks than other sites along the east coast of North America. Are there particular faults or other factors that make the sites of the shocks in 1886 and 1929 more active long term? We don't know. An answer to this question is very relevant when assessing earthquake risks for other populated parts of the coast, such as Boston, Providence, and New York City.

Several large and damaging historic earthquakes have occurred along the St. Lawrence River to the north of Maine (figure 11.1). That seismic zone is distinct from the one that extends northwesterly into Canada from northern New York State. Shocks of about magnitude 6.5 to 7.0 that occurred at a hot spot of activity just to the north of northern Maine in 1663, 1870, and 1925 were felt over huge areas of eastern Canada and the United States. Those earthquakes were concentrated along more numerous faults that were formed by an ancient meteorite-impact structure.

Figure 11.1 indicates a zone of well-located, recent activity in southern New Hampshire and northeastern Massachusetts, which was the site of several moderate-size earthquakes, including an event of magnitude 6.0 off Cape Ann, Massachusetts, in 1755. That shock has been studied and much debated because it was the largest earthquake used in the design of the nearby nuclear-power reactors at Seabrook along the coast of New Hampshire. I think that magnitude 6.0 shocks, like the one near Cape Ann, are possible as far south as New York City.

Three major earthquakes occurred during 1811 and 1812 in southeastern Missouri, northeastern Arkansas, and western Tennessee in an area that is called the "New Madrid seismic zone." New Madrid was a tiny frontier town on the Mississippi River in 1811. The consensus today is that the largest event was about magnitude 7.5 and the other two about 7.25. Previous estimates of their magnitudes were closer to 8.0. More recent shocks shown in figure 11.4 outline that active zone and may represent aftershocks of the 1811 and 1812 earthquakes.

Those three large shocks were felt over huge areas of the central and eastern United States, areas about five to ten times larger than the felt area for shocks of the same magnitude to the west of the Rocky Mountains. Earthquakes of those large magnitudes do occur in the central and eastern United States but about five times less frequently than those in the West. In 1990, Stuart Nishenko of USGS and G. A. Bollinger of Virginia Tech found that because large shocks in the central and eastern United States are felt to much greater distances, they are only one-third less likely to produce an earthquake with a comparable area of damage and societal impact in the coming 30 years than are shocks in California. They emphasized that the response of buildings and critical structures to a large, damaging earthquake within or adjacent to the New Madrid zone has not been tested since the large shock of 1895. Many people live in that broad area of potential damage.

Earthquakes of the New Madrid zone extend into southern Illinois and thence to the north along the border of Indiana and Illinois. Trenching (excavation) to investigate a fault along that zone in the Wabash Valley identified a large prehistoric earthquake. Data from this earthquake and the nineteenth-century New Madrid shocks have been used in the design of reactors and other critical structures in the central United States.

Another prominent zone of earthquakes shown in figure 11.4 extends from southwestern Virginia across eastern Tennessee and into northern Alabama. It is the site of the Virginia earthquake of 1897. The Mineral, Virginia, earthquake of 2011, magnitude 5.8, occurred within another distinct seismic zone in central Virginia and was comparable in size to the event in 1897.

Compressive Stress in the Eastern and Central United States

In 1972, Lamont graduate student Marc Sbar and I collected information on compressive stresses (pressures) in the crust of the earth and their bearing on the occurrence of earthquakes in eastern North America. We found that many published estimates of stress in that region were made close to large openings in mines to ascertain the safety of shafts and tunnels, so they were not appropriate for our goal of obtaining measurements that reflected the state of stress in the upper few miles (kilometers) of the earth's crust.

We did find some published measurements that were appropriate to understanding crustal stresses. One technique used to derive these measurements is called *overcoring*, in which a hole is drilled into hard rocks far from a tunnel in a mine. The change in shape of the drill hole over time yields a measurement of stress. Another technique called *hydrofracturing* involves making pressure measurements in boreholes up to a mile (1.6 kilometers) deep in the earth. Water pressure at a chosen depth is increased until the surrounding rock fractures in a direction that is then measured. Commercial fracking of sediments to obtain oil and gas is an extension of the hydrofracturing technique.

Sbar and I also obtained directions of maximum compression from mechanisms of earthquakes. Thrust faulting since the most recent glaciation generated young features called *pop-ups*, where rocks are deformed upward at the earth's surface. We used the orientation of those features as stress indicators. Some pop-ups were also formed soon after rock was removed in some quarries in western New York. A similar example occurred within a quarry near Wappinger Falls, about 50 miles (80 kilometers) north of New York City, where removal of rock decreased the vertical stress, causing sudden failure and hence an earthquake. Failure would not have occurred unless one of the horizontal stresses was very large and compressional.

In solid materials such as the earth's crust, stress, a measure of force per unit area, often differs considerably in each of three perpendicular directions. In liquids, however, stress (pressure) is the same in all directions. If stresses in crustal rocks were generated only by gravitational acceleration, the vertical stress would be about three times larger than the horizontal stress. The vertical stress at a given depth is the weight per unit area of the rocks above that point. Vertical stress about three times larger than horizontal stress is found in barely consolidated young rocks along the coast of the Gulf of Mexico.

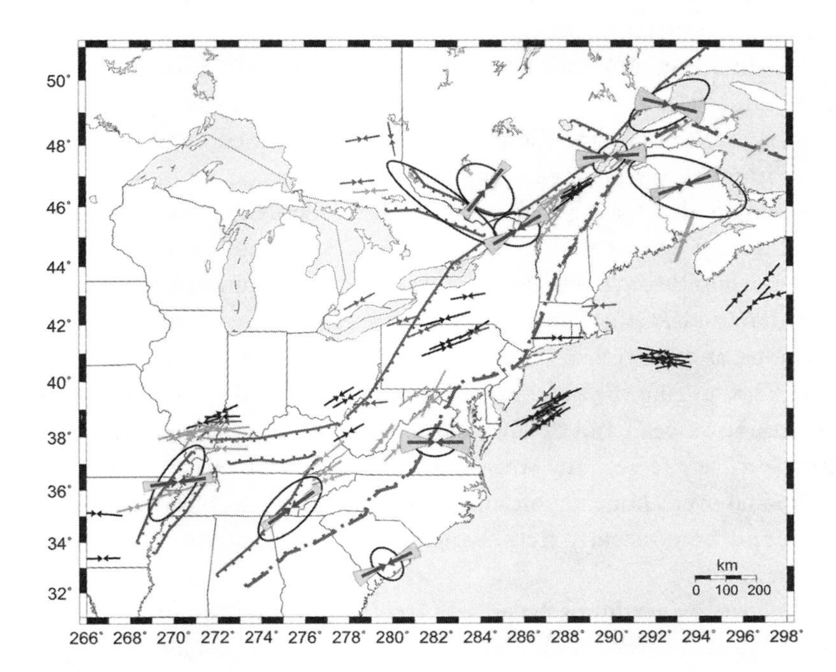

FIGURE 11.5 Directions of compressive stress in the eastern United States and adjacent parts of Canada. Shaded arrows indicate directions from earthquake mechanisms. Other arrows denote measurements in boreholes. See plate 25.

Source: Mazzotti and Townsend 2010.

In much of the eastern and central United States, however, Sbar and I found that one of the horizontal stresses in older rocks is much larger than the vertical stress. Hence, the state of stress in these areas involves more than just the weight of the rocks above a certain depth. Figure 11.5, a recent compilation, indicates that the directions of maximum compressive stress in huge areas of the United States and Canada are oriented northeast to east, with the exception of one zone along the St. Lawrence Valley. Within each subarea, the measured directions are quite consistent.

The directions of maximum compression have a great bearing on which faults in a region are seismically active today. For example, earthquakes occur on thrust faults oriented northwesterly in northern New York and eastern Canada in response to maximum compressive stresses oriented perpendicular to those faults—that is, northeasterly. In those areas, the smallest compressive stress is vertical. Strike-slip faulting occurs along the Dobbs Ferry fault in southern New York in response to maximum compression oriented east–northeast.

Maximum compressive stress in the New Madrid fault zone is also oriented east–northeast nearly perpendicular to the geological feature known as the Mississippi Embayment. The Embayment is an indentation of a former shoreline opened previously by extension. In other words, it was being pulled apart horizontally, not by compression, around 150 to 40 million years ago. Several igneous rock bodies were injected into the crust from the mantle during that period, including ones in Arkansas that contain diamonds. Other igneous rocks were emplaced in New Hampshire, Vermont, and Virginia as well as offshore of New Jersey and near Ithaca, New York, during the same time interval, well after the initial opening of the Atlantic Ocean and the formation of the extensional Newark basin in northern New Jersey. The stress field changed from extensional to compressional over a huge region sometime later and is the one present today. Why and how recently that change occurred are questions of current research.

Because knowledge of the present stress field is very important to many problems of earthquake risk, an understanding of its generation begs for continued research. Although we can map the directions of compressive stress, we know little about how big the stresses are or how they developed.

Earthquakes Triggered by Fluid Injection

Some earthquakes in the central and eastern United States have been triggered by the injection of waste fluids into deep disposal wells. A good example is the earthquakes triggered in the 1960s by the injection of waste liquids from the manufacture of nerve gases at the U.S. Army's Rocky Mountain Arsenal just to the northeast of Denver, Colorado. Those shocks increased in size with time to magnitude 4.85. Some caused damage in a region of otherwise very low natural seismic activity. When the likely cause of the Denver shocks was identified, pumping was halted in early 1966, and the earthquakes ceased the next year.

The injection of fluids at high pressure into rocks that were already under high natural stresses reduced the resistance to fault motion. It triggered the Denver earthquakes along what had been old, inactive faults in hard rock. The earthquakes were a major reason Denver was not able to compete for a planned large-particle accelerator, which instead went to Illinois.

My Lamont colleague Won-Young Kim studied a series of earthquakes near Youngstown, Ohio, that were triggered in 2011 and 2012 by the disposal of waste fluids in a deep well. No known natural earthquakes had been identified previously in the area. The largest of the series, magnitude 3.9, was widely felt. The series occurred along a fault that was oriented like the Dobbs Ferry fault in southern New York in 1985. Thus, the orientations of the maximum compressive stresses in the two areas were similar.

In parts of Oklahoma, earthquake activity recently increased dramatically following the disposal of waste fluids into disposal wells. A shock of magnitude 5.6 near Prague, Oklahoma, in 2011, caused considerable damage. It was the largest known earthquake in Oklahoma. Parts of Oklahoma now experience more small and moderate-size earthquakes than California.

The disposal of large amounts of fluid waste in disposal wells caused much of the increase in earthquake activity in Oklahoma, Texas, Arkansas, and Colorado during the past several years. Those wells typically received wastes from wells in the region in which hydrofracturing (fracking) was used to free either oil or gas from so-called tight geological formations in which petroleum would not flow otherwise into wells. Fracking has led to a great increase in the production of gas in several states of the central and eastern United States and to increased oil extraction in North Dakota and eastern Montana. The disposal well near Youngstown, Ohio, has received waste fluids from wells in adjacent Pennsylvania that were hydrofractured for gas extraction.

Disposal wells typically operate for many years. Although thousands are in use today in the United States, relatively few are known to have triggered earthquakes. When they do, however, the causal connection is understandably of great public and governmental concern. In contrast, wells in which hydrofracturing is conducted very rarely trigger even small earthquakes. Hydrofracturing is a relatively short operation and does not involve the larger amounts of fluids often injected into disposal wells.

Much research is currently under way to ascertain beforehand whether a site is likely to trigger earthquakes once a disposal well is used. Some of the factors that appear more likely to trigger earthquakes are high injection pressures, large volumes of fluids, injections into old faults, high natural stresses, and strong rocks at the disposal point. Proposals to dispose of high-level nuclear wastes in deep wells in the crust have been revived since the plan to use the underground facility at Yucca Mountain, Nevada, was halted. The triggering of earthquakes and the dangers of injecting toxic materials into the earth are likely to keep underground waste disposal an ongoing issue.

12

EARTHQUAKE RISKS TO NUCLEAR-POWER REACTORS

The U.S. Atomic Energy Commission (AEC) was responsible for the approval and licensing of the first reactors for electric-power generation in the United States in the 1960s and early 1970s. In 1974, the AEC was split into two parts. One, now called the Department of Energy, became responsible for developing nuclear weapons and promoting nuclear power. The Nuclear Regulatory Commission (NRC), a new federal agency, became responsible for licensing nuclear-power plants and ensuring their continued safety.

By 2014, one hundred nuclear power plants in the United States had been licensed to operate at sixty-five sites in thirty-one states. They produce about 20 percent of the electricity used in the country. Most were and still are located east of the Rocky Mountains.

The energy liberated in nuclear fission is huge. It is much greater than the energy released in ordinary chemical reactions, which merely involve the lightweight electrons that surround the nucleus. Today each large U.S. power reactor runs on about 100 tons of fuel enriched to 4 to 5 percent uranium-235. The fissioning, or splitting, of the nucleus of atoms of uranium 235 produces heat, which in turn is used to generate electricity. U.S. power reactors also produce about 1,500 tons of highly radioactive fission products and 400 tons of plutonium per year. Many of these reactors, initially licensed for 40 years, have already operated near that limit. There is considerable concern about whether they can continue to operate safely if they receive license extensions for additional periods of 20 and even 40 years, particularly if they are not upgraded.

Many of the personnel who moved to the NRC from the AEC strongly supported nuclear power and favored the nuclear industry. Both the AEC and the NRC frequently denied allegations about poor reactor design and safety from those who challenged these bodies' regulations. The NRC's inspector general observed that the NRC appeared to have informally established an unreasonably high burden of absolute proof on those contesting safety issues. I think that insistence on a high burden of proof continues today.

Relatively few geological and seismological studies were done initially for most of the early reactor sites in the United States. Utility companies proposed two sites in California—Santa Monica on the west side of the Los Angeles metro area and Bolinas on the San Andreas Fault to the northwest of San Francisco. At first, these sites were considered desirable because short transmission lines could bring electricity to nearby cities. A belief at the time was that engineers could design a reactor that would survive a nearby large earthquake. Nevertheless, these sites were ultimately rejected as being too close to major cities.

Indian Point

In 1962, the AEC licensed an electric-power reactor called Indian Point 1 about 35 miles (56 kilometers) north of midtown New York City. Planning for Indian Point 1 commenced in the late 1950s. I became involved with the earthquake safety of two additional much larger reactors at the site—Indian Points 2 and 3—in 1974 (figure 12.1) soon after we expanded the Lamont network of regional seismic-monitoring stations. In 1976, I testified before the U.S. NRC Atomic Safety and Licensing Appeal Board about whether the reactors at Indian Point, New York, were designed well enough to resist strong shaking.

Indian Point is located on the Hudson River 19 miles (31 kilometers) north of my home in Palisades and within 50 miles (80 kilometers) of more than 17 million people, including people in parts of Connecticut and New Jersey as well as in southern New York State. No matter what distance is

FIGURE 12.1 Indian Point nuclear-power plants, Buchanan, New York, looking east across the Hudson River. See plate 26.

Source: Photograph courtesy of Tony Fisher, 2008, Wikimedia Commons.

picked between 10 and 75 miles (16 and 120 kilometers), Indian Point is located closer to more people than any other reactor in the country.

In planning for a major accident, the NRC has emphasized only distances within 10 miles (16 kilometers) of reactors. This radius, however, was not adequate in Japan after the Fukushima accident in 2011, where significant doses of radiation deemed unsafe to humans extended well beyond 19 miles (30 kilometers). At the height of the accident, the United States advised its citizens living there to move 50 miles (80 kilometers) from the Fukushima reactors.

Fortunately, winds initially carried much of the radioactive debris from Fukushima out to sea for several days. In a major disaster at Indian Point, however, winds would carry high-level radioactive debris long distances over land no matter their direction at the time. The most severe consequences would happen if winds blew toward highly populated areas either to the southeast or south to New York City. Even though most roads within 10 miles (16 kilometers) of Indian Point are narrow and winding (thus making the site difficult to access), the NRC nonetheless stated in 2002 that off-site emergency plans are adequate to protect public health and safety. The word *adequate* appears in many NRC judgments and pronouncements.

In testimony before Congress in 2002, Richard Meserve, chairman of the NRC, said, "Before September 11, 2001, nuclear power plants were among the best defended and most hardened facilities of the Nation's critical infrastructure." Many scientists and engineers disagree with this statement and similar statements by others.

The only investigations of earthquake risk done before Indian Point 1 was licensed in 1962 appear to be several hours of consulting by Father J. Joseph Lynch of Fordham University, a Jesuit priest and seismologist. He wrote one of the reports on Indian Point 1, saying, "The probability of a serious shock occurring in this area for the next several hundred years is practically nill [*sic*]. The area therefore would certainly seem to be as safe as any area at present known." He also claimed, "Estimated maximum ground acceleration of 0.03 g is reasonably conservative for the area." Lynch's opinion seems to have been based on large earthquakes in very active areas such as Japan and western South America and on lesser activity in other areas. As a consequence, Indian Point 1 was constructed to withstand only a small earthquake.

In 1974, Indian Point 1 was closed because it didn't have an emergency core-cooling capability. Some of its nonnuclear components, such as a tall chimney, continued in use for decades afterward for later Indian Point reactors.

Disaster specialists distinguish between earthquake hazards and earthquake risk. The term *earthquake hazard* refers solely to the amount of shaking in an earthquake at a given locality. Thus, the hazard for New York City has not changed over time since Henry Hudson sailed up the river in 1609. The term *earthquake risk* refers to hazard times the assets or people exposed times their vulnerability or fragility. Earthquake risk has increased greatly over the past 400 years. Whereas earthquake hazard in the New York City area is relatively low, earthquake risk is high once the huge numbers of people, buildings, and infrastructure are factored in. The NRC pays much more attention to hazard than to risk, a grave mistake for critical facilities such as reactors.

Indian Point Hearings Before the NRC

In 1973, James Davis, the New York State geologist, hired my former Lamont colleague Paul Pomeroy as state seismologist. At the same time, a new and much larger reactor, Indian Point 2, was nearing completion, and the foundation for reactor 3 was under construction. Pomeroy and Davis became concerned that not enough work had been done on earthquake safety for the new reactors (Davis 1974).

My sense was that Consolidated Edison of New York (Con Edison), the owners of the two new reactors, paid little attention to earthquake risk prior to the time these two New York State employees raised concerns. The state negotiated with Con Edison to have a seismic network installed around Indian Point and for geologist Nicholas Ratcliff, a professor at the City College of New York, to conduct detailed mapping of faults in the vicinity. In his report on faults of various ages, Ratcliff identified a number of faults that had not been healed by a heating (metamorphic) event. It was not possible to tell, however, whether they had undergone displacements recently or if they had been inactive for much of the past 175 million years. One of those faults ran under the foundation for Indian Point 3, a finding that led New York State to be more concerned about earthquake safety.

Indian Point, in fact, is located less than one mile (1.5 kilometers) from the Ramapo fault, which forms the boundary between the Newark basin and the older rocks of the Reading Prong that stretches from Reading, Pennsylvania, through northern New Jersey and southern New York and terminates in western Connecticut (figure 11.2). Pomeroy assembled reports of shaking from a magnitude 3.85 earthquake in 1951 and found they centered on the Ramapo fault not far from Indian Point. His study (see Davis 1974) led me to work on earthquakes near the Ramapo fault as well as on others in the greater New York City area.

New York State brought a legal injunction against Con Edison regarding the construction of plants 2 and 3, claiming that not enough geological and seismological work had been done according to NRC regulations to ascertain if the Indian Point reactors were safe.

New York's claims were heard in 1976 and 1977 by an NRC Appeal Board, formal legal proceedings with lawyers representing each party. Various technical experts, including me, submitted written testimony, were questioned by the three members of the Appeal Board, and were cross-examined by lawyers for the various parties to the hearings—originally the State of New York, the NRC staff, Con Edison, and a local environmental group. Midway through the hearings, the Power Authority of the State of New York also became a party when it purchased Indian Point 3 from Con Edison for almost $1 billion. This purchase left the state with a conflict of interest on the issues at the hearings. The ruling by the U.S. NRC Atomic

Safety and Licensing Appeal Board (1977) contains a summary of various arguments.

The NRC Appeal Board and the lawyers for all parties agreed to divide the Indian Point seismic hearings into four issues. The legal basis for the hearings was "Appendix A: Seismic and Geologic Siting Criteria for Nuclear Power Plants," which the AEC added in 1973 to Regulations Title 10, Code of Federal Regulations, Part 100, "Reactor Site Criteria" (U.S. NRC 1973).

This regulation concerns the location of a reactor and its proximity to faults and historic earthquakes. In 1974, Appendix A became an NRC statute. Indian Point 1, however, was licensed before the regulation was put into effect.

Appendix A defined a "safe-shutdown earthquake" (SSE) as one based on an evaluation of the maximum earthquake potential for shaking at a reactor site. The SSE is also called the "design-basis earthquake." Electrical generation at a plant was to be shutdown following an SSE so that the condition of the reactor could be reassessed before USNRC permitted it to be restarted. A reactor was allowed to continue operating following a second somewhat smaller shock called the "Operating Basis Earthquake."

Design Earthquake for Indian Point

Issue one at the hearings involved the following question: "Under Appendix A does the Cape Ann, Massachusetts earthquake of 1755, or any other historic event, require a Modified Mercalli (MM) intensity greater than VII for the Safe Shutdown Earthquake at the Indian Point site?" The MM scale extends from intensity I (felt only by a few people) to intensity XII (total damage). Reports of intensities—that is, vibratory motion of the ground—were used because data on such reports existed for historic earthquakes well before magnitudes could be calculated from seismic records. Historic records extend back about 300 years for the northeastern and mid-Atlantic United States and back about 200 years for the rest of the forty-eight contiguous states.

The media and others often confuse magnitudes and intensities. *Magnitude* measures the energy released at the source of the earthquake based on seismograph recordings. *Intensity* measures qualitatively the strength of shaking produced by the earthquake at a given place and is determined by its effect on people, structures, and the environment. It should be remembered, however, that information from older historic shocks is often scarce. An intensity of MM VII corresponds to a magnitude of about 5.1, and an intensity of MM VIII to a magnitude of about 5.8.

Consultants for Con Edison and the NRC scientific staff claimed that a design basis of MM VII, which involves *negligible* damage to buildings of good design and construction and *slight to moderate* damage for well-built ordinary structures, was more than sufficient for the Indian Point reactors. New York State maintained, however, that the Indian Point reactors should be designed for intensity MM VIII, which involves *slight* damage to specially designed structures and *considerable* damage to ordinary substantial buildings, some with partial collapse. MM VIII includes the fall of chimneys, columns, monuments, and walls. Clearly, MM VIII involves greater damage than MM VII. For a shock of a given magnitude, intensity understandably diminishes with distance from its epicenter as well as with increasing depth of its hypocenter or focus. Earthquakes located deeper in the earth are not felt as strongly as those located closer to the surface.

All parties to the hearings accepted that shaking and damage in the Cape Ann earthquake of 1755 off the coast of Massachusetts and New Hampshire was of intensity MM VIII. Shaking and damage in each of the three great New Madrid, Missouri, shocks of 1811 and 1812 were rated as at intensity MM XII; in the earthquake in Charleston, South Carolina, in 1886 as MM IX to MM X; and in several earthquakes in the St. Lawrence Valley of Quebec as MM VIII to X. Earthquakes near Newbury, Massachusetts, in 1727, East Haddam, Connecticut, in 1791, and Cornwall, New York, in 1944 were rated as having an intensity of MM VIII.

Appendix A states that the largest earthquakes within 200 miles (320 kilometers) of a reactor site must be considered in its design. Each shock is to be positioned at the closest point in its *tectonic province* to the proposed reactor site and the MM intensity calculated for it at the location of a reactor. The largest value of MM intensity is to be used then for the design of the reactor. Appendix A of 1973, which was in use at the time of the hearings in 1976 and 1977, defines a tectonic province as "a region of the North American continent characterized by a relative consistency of the geologic structural features contained therein," a fairly vague definition. New York State claimed that the Cape Ann shock of 1755 and several others of intensity MM VIII occurred in the same large tectonic province as Indian Point, which it called the "Folded Appalachian" province. Hence, the state argued that intensity MM VIII was the appropriate design intensity for Indian Point. Several of us who testified on behalf of New York State said that not enough was known about tectonic provinces as *they relate to the occurrence of earthquakes* and thus questioned the validity of subdividing the region within 200 miles (320 kilometers) of Indian Point into separate smaller tectonic provinces.

Nevertheless, consultants for Con Edison and the NRC staff argued that a number of tectonic provinces existed within 200 miles (320 kilometers) of Indian Point. The purported characteristics of most of these provinces, however, were based on mountain-building events that occurred more than 350 million years ago. Con Edison, the NRC, and the State of New York also defined an additional separate coastal plain province. What any of those provinces have to do with earthquakes and stresses in the crust today was not clear in 1976 and still is not today.

Scientists at Dames & Moore, Inc., a large civil engineering consulting firm hired by Con Edison, claimed that Indian Point was located in the "Conestoga Valley tectonic province." It was a baffling, unexpected designation that shocked most of us who were scientific witnesses because that valley is located in eastern Pennsylvania. Their purpose obviously was to use the largest known historic earthquake in that province, each with an intensity less than MM VII, as a guide in determining the earthquake design for Indian Point. They confined all other large earthquakes to other tectonic provinces and at large distances from Indian Point. Although the NRC staff did not place Indian Point in that Conestoga Valley province, they did delineate a number of different tectonic provinces within 200 miles (320 kilometers) of Indian Point. By their logic, there were no earthquakes of intensities greater than MM VII in Indian Point's province. They assigned the earthquake of intensity MM VII just offshore of New York City in 1884 to a coastal plain tectonic province. When moved to the closest point in that province, that earthquake was 45 miles (72 kilometers) from Indian Point. Its calculated intensity at the reactors was therefore much less than MM VII.

I considered this division into many tectonic provinces a "gerrymandering" of significant earthquakes within 200 miles (320 kilometers) of Indian Point. (In the processes of setting electoral voting districts, gerrymandering establishes a partisan political advantage by manipulating the boundaries of districts.) This "gerrymandering" purposely reduced the intensity of the reactor's design-basis earthquake. The NRC staff never presented a map illustrating the location of the tectonic provinces they used at the hearings. In addition, they employed yet a different set of tectonic provinces for the design of several other reactors in the northeastern United States.

Two of the three members of the Appeal Board, John Buck and Lawrence Quarles, wrote the majority opinion for the maximum-intensity issue. They stated, "We must conclude that the approach taken by the licensees [Con Edison] in formulating their provinces is the correct one." They said the licensees made the only consistent attempt to utilize a range of scientific data in their interpretation of Appendix A. They went on to say that a major basis for that approach was the present tectonic plate model of past movements of the American and African continents, which their witnesses considered to represent the current "state of the art." Buck and Quarles commented, "State witness [James] Davis branded the use of the tectonic plate theory to explain the deformation of the Appalachians as supposition and speculation" (U.S. NRC Atomic Safety and Licensing Appeal Board 1977).

In the ruling handed down in 1977, Buck and Quarles indicated they were very impressed by testimony of Con Edison's consultants on plate tectonics. They completely misinterpreted the testimony of New York State chief geologist, James Davis, who cast doubt neither on the tenets of plate tectonics nor on the formation of the Appalachians. Davis merely doubted that the region should be subdivided into many tectonic provinces *for the purpose of assessing earthquake risk.* Buck and Quarles said, "We find nothing in Appendix A that prohibits us from utilizing the latest accepted geologic developments in making determinations of tectonic provinces." It is quite possible that Buck and Quarles heard about plate tectonics for the first time at the hearings and knew nothing about my being involved as one of its main developers.

Clearly, the definition of tectonic provinces by consultants for Con Edison did not constitute "latest accepted geologic developments." The USGS, which had a major role in developing Appendix A, had expressed the opinion that "the Appendix would be extremely difficult to apply and would lead to a lot of confusion in the assigning of tectonic provinces."

The chair of the Appeal Board, William Farrar, issued dissenting opinions on three of the four seismic issues. His short opinion is attached at the end of the majority opinion issued in 1977. Farrar dissented on seismic issues for Seabrook and Indian Point in a longer opinion he released later. He said,

> A number of earthquakes greater than Intensity VII . . . have occurred in and around the Eastern seaboard in the past 200 years. Under the [Nuclear Regulatory] Commission's regulations, we must assume that earthquakes of this size will recur; the question is—where? . . . In deciding which of the widely varying versions of proposed tectonic provinces to accept, we must be conscious of the context of our inquiry. In searching for "relative consistency," we should be paying particular attention to those features which are similar or dissimilar in terms of what they signify in terms of earthquake potential. In other words, structural differences which have no discernible bearing on the present likelihood of earthquakes should not, as I read the regulations, form the basis for drawing province boundaries. (Farrar 1979)

That was and still is my opinion as well.

The majority opinion left the design intensity for Indian Point at MM VII. Has that design intensity held up in the decades since the hearings? A key factor in the generation of earthquakes is the presence of high compressive stresses in the earth's crust, which are quite uniform in direction throughout New England and the mid-Atlantic states (figure 11.5). The generation of earthquakes in this intraplate region requires faults that have not been healed by heating events and ones that are oriented favorably for failure within the present stress configuration. Because the stress field changed sometime in the past 40 to 90 million years, earthquakes today likely are not controlled by older stresses but by present ones. This situation argues for one large tectonic province and that MM VII was too small for the design-basis earthquake used in developing Indian Point.

Converting Intensities Into Ground Accelerations for Reactor Design—North Anna and the Virginia Earthquake of 2011

The second issue argued in 1976 was "Should the ground acceleration value [at the plants] used for the design of Indian Point be increased?" This issue involved converting intensity of seismic shaking in MM units at a site into the engineering design for a reactor. Designs for Indian Point and other reactors in the United States were summarized in graphs of expected seismic acceleration of ground shaking. I use the engineering graph for the two reactors at North Anna, Virginia, in figure 12.2 to illustrate a number of points because these reactors were shaken beyond their designated "safe-shutdown earthquake" by the nearby shock of magnitude 5.8 in Mineral, Virginia, on August 2011. They are similar in design to reactors 2 and 3 at Indian Point.

Figure 12.2 indicates shaking of the earth's surface at the North Anna reactors during the 2011 earthquake. The black line was chosen as the design for the safe-shutdown earthquake (SSE). The observed north–south (NS) shaking during the earthquake in 2011, shown by the lightest-gray line, was larger than the observed east–west (EW) shaking. Hence, I pay attention here to the north–south observation.

The SSE design (in black) is somewhat larger than the observed value (light gray) at the left of figure 12.2 but only for low frequencies on the horizontal axis from 0.25 to about 2 cycles per second (Hertz). The lowest

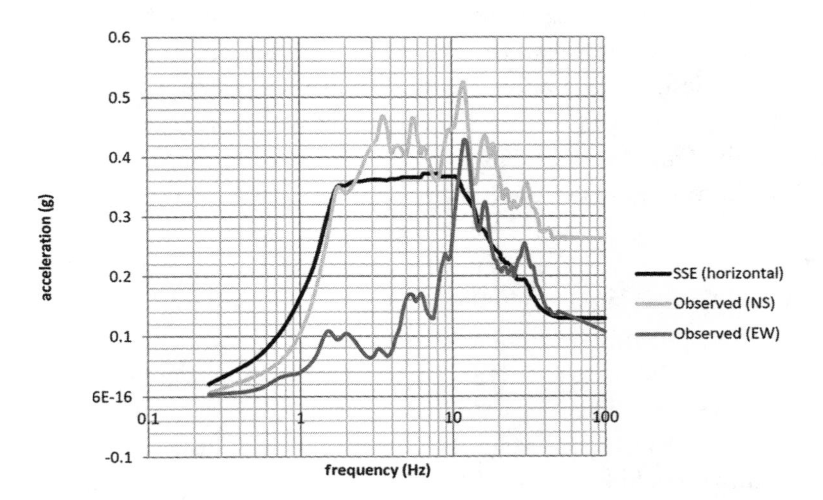

FIGURE 12.2 Comparison of observed horizontal accelerations of the surface of the earth during the Virginia earthquake of 2011 as a function of frequency. Calculated values for the safe-shutdown earthquake (SSE) at North Anna are indicated by the darkest line. See plate 27.

Source: U.S. NRC 2011.

(fundamental) frequency of the main containment structures at North Anna and Indian Point is larger, several Hertz—that is, several cycles a second. The design SSE performed well for the earthquake of 2011 up to 2 Hertz but not for higher frequencies, where the light-gray observed curve is well above the calculated SSE in black. (In comparison, electrical current in homes in the United States alternates at 60 Hertz—that is, sixty times a second.)

The observed acceleration during the Virginia earthquake was one and a half times higher at a frequency of 12 Hertz and two times higher at 40 Hertz than the design. The higher design accelerations between 2 and 10 Hertz in figure 12.2 are associated with shaking near the fundamental frequencies of the reactors. Shaking exceeding certain accelerations is referred to as being "beyond the design basis" for reactors. Strong debate continues about the safety of reactors for which the SSE is too small. In 2016, the NRC finally ruled that beyond design-basis shaking must be considered in assessing the safety of older U.S. reactors.

North Anna was designed for a ground acceleration of 0.13 g for the high frequencies, shown at the far right side of figure 12.2. (Gravitational acceleration at the surface of the earth is 1.0 g.) The value at the far right side, which is called the "peak ground acceleration," or PGA, is confusing because it is not the frequency with the highest acceleration on the graph but one where the design acceleration flattens out at very high frequencies (Hertz).

The design PGA of 0.15 g for Indian Point was nearly the same as that for North Anna. At the Indian Point hearings in 1976, consultants for Con Edison argued that a PGA of 0.13 g was associated with shaking of MM VII and hence that the design PGA of 0.15 g was more than adequate. The design PGA is five times stronger than what Lynch claimed was reasonably conservative for Indian Point 1 in the 1960s—that is, Lynch's value was far too small for even modest amounts of shaking.

Buck and Quarles of the NRC Appeal Board accepted a PGA of 0.13 g in 1977 over the dissent of Chairman Farrar. Farrar said, "But I cannot agree that .13 g has been shown to represent the maximum effective acceleration that could be expected to result from such an earthquake. Thus, whether the plant's .15g design is adequate deserves at most the Scotch verdict 'not proven' " (Farrar 1979).

Mihailo Trifunac, an expert on seismic engineering and strong shaking at USC, argued before the Appeal Board in 1976 that a larger PGA should be used because pertinent data did not exist for the central and eastern United States. He stated that shaking in those areas would be larger than shaking from earthquakes in the western United States, where some records of strong shaking existed in 1976. Large earthquakes in the eastern and central United States are felt and cause damage at much larger distances than in the West. Strong shaking also occurs at higher frequencies in the eastern and central United States, as is shown clearly in figure 12.2 for the North Anna reactors. Trifunac concluded in 1976 that these factors led to design accelerations for Indian Point that were too small. Newer recordings of strong shaking that now exist for earthquakes in the eastern and central United States strongly support his conclusion of 1976.

The 5.8 magnitude of the Mineral, Virginia, earthquake in 2011 was greater than the 4.8 magnitude of the largest known historical earthquake in the central Virginia seismic zone. Its occurrence highlights the need to account for earthquakes of larger size than only those of known historic shocks when constructing critical structures such as nuclear-power plants. Nevertheless, the NRC's Appendix A of 1973 stipulates using just known historic earthquakes. This was one of several shortcomings of Appendix A as applied to reactor sites east of the Rocky Mountains.

The USGS publishes probabilities of shaking in large earthquakes that will have only a 2 percent likelihood of occurring at various locations in the United States within the next 50 years. To make those calculations, historical data must be extrapolated back 2,500 years. We have known for decades how to extrapolate the rate of occurrence of earthquakes from 300 to 2,500 years. The rate of occurrence of shocks falls off at a known rate with increasing seismic magnitude—that is, earthquakes of magnitude 6.0 occur in a region about 5 to 10 percent as often as earthquakes of magnitude 5.0.

A 2 percent chance that shaking will exceed a certain amount during the next 50 years is often used today for critical structures. Most of the nuclear-power reactors in the United States, however, were not designed for shocks larger than those in historical records. For noncritical structures, USGS also calculates earthquake magnitudes for which the chance that shaking occurring at various places will have less than a 10 percent (not 2 percent) chance of occurring during the next 50 years. Those estimates involve extrapolating data back 500 years, a period still longer than historical U.S. records. Clearly, it would not be as safe to have many reactors in the United States designed for only a 10 percent chance of exceeding their design in 50 years as opposed to a 2 percent chance.

The central Virginia earthquake of 2011 damaged the Washington Monument and parts of the National Cathedral in the District of Columbia 80 miles (130 kilometers) away. Strong shaking even in the New York City area during that earthquake led to closure for hours of an air-traffic-control facility. Although many people and some scientists were surprised that damage occurred far away, they should not have been.

Several years ago I listened by phone to an open discussion at an NRC meeting about earthquake design. A member of the NRC staff argued that shaking greater than three cycles per second (3 Hertz) is not significant for nuclear-power reactors. Nevertheless, a reactor is a complex system whose components are sensitive to many different frequencies. Several years ago engineers at Westinghouse, a designer of reactors, paid great attention to frequencies much higher than 3 Hertz in designs for new reactors because they considered those frequencies important in resisting strong shaking. By 2007, the NRC and its consultants were also paying greater attention to frequencies as high as 25 and 50 Hertz—that is, frequencies at which shaking in the Virginia earthquake exceeded the design basis for the reactors at North Anna (figure 12.2).

Thus, a number of lines of evidence indicate that the design accelerations for frequencies higher than 2 to 3 Hertz for Indian Point and North Anna were too low. Specifying designs to include only historic shocks was a poor choice.

Active (Capable) Faults

The third issue that was litigated about Indian Point in 1976 was "Is the Ramapo fault a capable fault within the meaning of Appendix A?" Because geological opinions varied about what constituted an active fault, the NRC followed the AEC in using the term *capable fault* from Appendix A of 1973. That term, however, is neither known nor used in the scientific literature except by the NRC. Thus far, neither the NRC nor its Appeal Boards have declared any faults near reactors in the central and eastern United States to be capable. They implicitly assume that faults in the vast area to the east of Salt Lake City Utah are not active enough to be taken into account in reactor design.

New York State seismologist Paul Pomeroy showed in 1974 that reports of shaking in an earthquake in 1951 were centered on the Ramapo fault (figure 11.2). At the hearings in 1976, I used this earthquake and the locations of several small shocks in my argument that the Ramapo fault should be considered a capable fault for the design of Indian Point. In the middle of the hearings, a well-located earthquake of magnitude 2.5 and an aftershock occurred on the Ramapo fault. More accurate travel times of seismic waves from this shock allowed me to compute a more accurate location for the earthquake in 1951. I found that it occurred about 6 miles (10 kilometers) northwest of the Ramapo fault, within what we later called the "Ramapo seismic zone."

Con Edison was shocked by its consultants' poor presentations in 1976 on the issue of tectonic provinces. For the issue of the Ramapo fault, the company brought in Charles Richter of Cal Tech to state that it should not be considered a capable fault. Appendix A gave the following vague criterion for designating a fault as capable: "Macroseismicity instrumentally determined with records of sufficient precision to demonstrate a direct relationship with the fault." Most seismologists rarely, if ever, use the word *macroseismicity*, another problem with the formulation and use of Appendix A.

During the hearings, Richter said, "I understand 'macro-seismicity' to refer to large and significant seismic activity like that observed in California, such as is generally associated with fault movement at the surface." When I assisted lawyer David Fleischaker in cross-examining Richter, it became clear that Richter had in mind earthquakes of either magnitude 7.0 or perhaps several events along a fault of magnitude 6.0 or greater. We planned to ask him more questions after lunch, but unfortunately the Appeal Board excused the 76-year-old Richter, saying he was tired. An unanswered question was: How do you evaluate shocks larger than the design earthquake for Indian Point, about 5.1, but smaller than the magnitudes discussed by Richter, 6.0 to 7.0?

The Appeal Board ruled in 1977 that the Ramapo fault was not a capable fault. In 2008, after 35 years of additional seismic monitoring, four of us at Lamont showed that a 7.5-mile-wide (12-kilometer-wide) zone of activity, the Ramapo seismic zone, extended to the northwest from the Ramapo fault into the hard rocks of the Reading Prong. A large earthquake either on the Ramapo fault or within that seismic zone would produce similar shaking at Indian Point.

In addition, we found that earthquakes of small and moderate size extend from the Manhattan Prong of New York City into the coastal plain just offshore of Brooklyn and New Jersey. That finding is not too surprising because hard, old continental rocks are found at shallow depths beneath those parts of the coastal plain. The young, soft rocks are merely the "icing" on top of the older rocks of the coastal plain, where earthquakes are more likely to occur. I find no justification for confining the 1884 earthquake solely to a coastal plain tectonic province for establishing the design-basis earthquake for Indian Point. Hard rocks beneath the coastal plain extend into adjacent hard rocks of the Manhattan Prong.

The Appeal Board heard arguments in 1977 about halting the network monitoring earthquake activity near Indian Point that had been supported by Con Edison. The NRC staff, New York State, and an environmental group argued for an expanded network. The NRC staff said, "We were not able to conclude conservatively that this structure [the Ramapo seismic zone] does not play a possible role in localizing earthquake activity" (U.S. NRC Atomic Safety and Licensing Appeal Board 1977). I did not participate in those hearings.

In 1977, Appeal Board members Buck and Quarles ruled against further monitoring, with Chairman Farrar dissenting. Indian Point was left without seismic monitoring except for a few Lamont stations at larger distances. The NRC staff concluded, and I agreed, that more attention should have been given to monitoring the higher activity along the Ramapo seismic zone. In 1997, the NRC revised Appendix A, changing the words *capable fault* to *capable tectonic source*. I think the Ramapo seismic zone can be called a capable tectonic source. Nevertheless, that expression is not used today except by the NRC.

Gaps in Understanding Earthquake Occurrence

In 1977, Chair Farrar said, "In recognition of the gaps in our understanding of earthquake occurrence and mechanism, the [Nuclear Regulatory]

Commission's regulations insist that in this area, more so than in others, conservatism be the watchword" (U.S. NRC Atomic Safety and Licensing Appeal Board 1977). I do not think that Buck and Quarles observed such conservatism, accepting the statements by Con Edison and their consultants as sufficient evidence.

In a longer dissenting opinion in 1979, Farrar said,

> The Commission's regulations seem to contemplate that clear findings will be at hand in precisely those areas where our knowledge is most limited. In the face of this uncertainty, I respectively suggest that my colleagues have neglected an elementary premise of nuclear power reactor regulation: to prevail, those who assert the adequacy of reactor design—not their opponents—must bear the burden of proof. I reiterate that, instead, they have view[ed] the evidence presented by the interveners with an unjustifiably jaundiced eye, demanding from them what they do not expect of the staff and applicants—strict proof neither within the grasp of any practitioners of the seismological arts nor demanded by the regulations. (Farrar 1979).

Debate about questions of the safety of U.S. reactors continues in chapter 14.

13

NUCLEAR-POWER REACTORS IN THE UNITED STATES

Lessons Learned from the Fukushima Disaster

Positions on Safety and Licensing of Reactors

I entered the Indian Point reactor hearings in 1976 prepared to testify about earthquake issues related to those reactors. With time, I became very concerned about the general disregard for design and safety of reactors by the NRC. The Fukushima nuclear disaster of 2011, described in chapter 10, contains lessons that can be learned and applied to reactors in the United States.

The NRC's rulings and letters to owners of reactors are full of jargon, abbreviations, and legal precedents. Ascertaining what are serious safety issues usually takes work and careful reading of the middle parts of its formal documents.

Challenging NRC regulations or rulings is very expensive and requires extensive legal preparation and representation. For example, in the hearings about earthquake design for Indian Point in 1976, the Citizens' Committee for Protection of the Environment spent about $4,000. I donated my services to the committee, as did its lawyer, for the issue of capable faults. New York State must have spent at least $1 million and Con Edison several times that. I decided to testify for that environmental group and for New York State because funding of the parties to the hearings was so unbalanced. The owners of reactors have spent huge amounts of money on consultants, whereas little funding has been available for others who have contested safety and design.

License Extensions

The NRC works in a strictly legal framework. It has turned down nearly all issues brought to it by interveners. Perhaps the best hope of obtaining action is through state governments, the political process, and the appointment of new NRC commissioners. Even the governors of New York and Vermont faced strong opposition from NRC and Entergy, the company that owns and operates several reactors, over the issue of 20-year license extensions. I worked with the Office of the Attorney General of New York in 2008 on two seismological contentions that were part of thirty-two challenges to the proposed 20-year extensions for the Indian Point reactors.

Nearly all of the nuclear-power reactors in the United States were designed and licensed several decades ago. Most of their owners have applied for 20-year extensions to the initial 40-year licenses. Until recently, the NRC commissioners ruled that the aging of reactors is the only issue than can be contested in granting license extensions. Disturbingly, neither the abundant new seismic information that has become available over the past decades nor the Fukushima accident, terrorism, or the great increase in the amount of spent fuel stored at U.S. reactor sites could be contested. The NRC commissioners ruled that new information must be incorporated into the licensing of the few new reactors proposed since 1997. Finally, in 2016, updating new findings for older reactors was required.

The NRC has granted 20-year license extensions to about 60 percent of existing reactors in the United States. The 40-year licenses for Indian Point 2 and 3 expired in 2013 and 2015. When litigation over 20-year license extensions dragged on, the NRC allowed Indian Point 2 and 3 to continue operating. Entergy, now the owner of Indian Point, has spent millions of dollars on public relations strongly favoring license extensions. It omitted the word *nuclear* in renaming the reactor complex the "Indian Point Energy Center." In early 2017, Entergy and New York State agreed that Indian Point reactors 2 and 3 would cease operations in 2020 and 2021. The governor of New York promoted that action.

Entergy, several public officials, and others who have favored continued operation of Indian Point have stated that New York City and Westchester County receive about 25 percent of their electricity from those reactors, so closing them would cause power outages and would increase the price of electricity. The 25 percent figure, which has been widely publicized, has not been correct for many years. Indian Point furnishes about only 5 percent of the electricity for New York City and adjacent Westchester County. Entergy sells the rest of its electricity from Indian Point to other parts of New York State and to New England.

The NRC as well as many owners of reactors argue that the safety of existing nuclear-power plants will not be compromised if licenses are extended. Some proponents even argue for additional 20-year extensions beyond 60 years of operation. Several reactors in the United States, however, either have closed or will close in the next few years for economic reasons, such as competition from natural gas and inexpensive oil. A major problem for those reactors that have closed is the cost of decommissioning them over the decades ahead and the final disposal of spent nuclear fuel.

Building a new reactor is estimated to cost $10 to $20 billion and to take five to ten years. As a consequence, few owners or lenders can afford the up-front costs of even one new reactor. Rather than building a new generation of safer reactors, the present direction is to continue relicensing older units. If a major accident happens to one of them, however, continuation of nuclear power in the United States likely would be jeopardized.

James Lee Witt's Assessment of a Major Nuclear Accident at Indian Point

After the disaster at the World Trade Center in 2001, New York State asked James Lee Witt and Associates, which specializes in emergency management, to assess response plans for a major nuclear accident at Indian Point. Witt was the director of FEMA from 1993 to 2001. The Witt Report of 2003 said the plants were built to comply with regulations rather than according to a strategy that leads to structures and systems that protect from exposure to radiation. It also stated:

> In our report we discuss significant planning inadequacies, parental behavior that would compromise school evacuation, difficulties in communications, outdated vulnerability assessment, use of outdated technologies, lack of first responder confidence in the plan, problems

> caused by spontaneous evacuation, the nature of the road system, the thin public education effort, and the impact of these on effective response in high population areas. None of these problems, when considered in isolation, precludes effective response. When considered together, however, *it is our conclusion that the current radiological response system and capabilities are not adequate to overcome their combined weight and protect the people from an unacceptable dose of radiation in the event of a release from Indian Point.* We believe a plant adjacent to high population areas should have different requirements than plants otherwise situated . . . The unique aspects of a terrorist event should not be dismissed by simply asserting that they are covered in current plans and exercises. (J. L. Witt Associates 2003)

Utilities sold many of the older nuclear-power plants in the United States to Exelon, Entergy, and a few other large corporations about 15 years ago, when electrical utilities were deregulated. Con Edison, like many utilities, stopped generating electricity and focused on the distribution of electricity. Entergy formed a limited liability corporation for Indian Point 2, another for reactor 3, and others for each of the nineteen reactors it owned elsewhere in the United States. It is quite possible that an individual limited liability corporation may not have the financial resources to deal with a major loss at its reactor. The liabilities in a major accident in the United States, like that at Fukushima, could well exceed several hundred billion dollars and perhaps even a trillion dollars.

Accidents at Nuclear-Power Plants in the United States

Several significant accidents have occurred at nuclear-power reactors in the United States. In 1975, workers using a candle to locate an air leak at the Brown's Ferry reactor in Alabama ignited a major fire that burned for seven hours and damaged more than 1,600 control cables. An accident in unit 2 of the Three Mile Island, Pennsylvania, plant in 1979 led to a partial core meltdown. Fortunately, its containment dome prevented the escape of high levels of radioactivity. Governmental agencies did not call for an evacuation, but many people left the area anyway.

Two of the five most severe accidents at power reactors in the United States since 1979 occurred at the Davis-Besse plants near Toledo, Ohio. In 2002, workers discovered that corrosion had eaten a large hole in one of the reactor vessels. The NRC imposed its largest fine to date—more than $5 millions—on the owners for actions that led to the corrosion. That fine, however, was a drop in the bucket because the facility generated that value of electricity in only a few days. The unit was shut for two years. Repairs and upgrades cost $600 million. The U.S. Department of Justice conducted a criminal investigation and fined the owners $28 million. Three persons were indicted for hiding evidence that leaking boric acid corroded the reactor vessel. Notwithstanding all of these problems, in 2010 the reactor's owner applied for a 20-year license extension.

Storage of High-Level, Spent Nuclear Fuel

Spent-fuel pools at U.S. reactors were intended originally for the temporary storage of highly radioactive spent (used) nuclear fuel after it was removed from reactors. Those pools were not built with containment structures like those enclosing reactor cores. Since the United States decided decades ago against reprocessing spent fuel, that fuel's remaining uranium could not be reused. This decision was made because it was felt that reprocessing might permit plutonium generated in the reactors to be diverted to nuclear weapons. Spent fuel in the United States was instead to be transported off-site for permanent storage for at least 10,000 years.

An underground facility to accept high-level spent fuel was constructed at great expense beneath Yucca Mountain in Nevada. Because the state government was not consulted, however, many Nevadans objected. In 2009, Senate majority leader Harry Reid of Nevada was able to halt work there. Since then, permanent disposal of high-level waste has been in limbo. Reid retired from the Senate in 2016, so the Yucca Mountain project may be reinstated by the Trump administration.

As a consequence of huge delays in creating a permanent storage site, most spent fuel in the United States continues to be stored at the sites of reactors, as it has been for several decades. The NRC now permits about five times more high-level radioactive waste to be stored in spent-fuel pools than its regulations originally permitted. At many individual U.S. reactors, the amount of spent fuel is greater than that stored at all of the reactors at Fukushima One. When either the level of water in spent-fuel pools becomes too low or the spent fuel is not cooled following an emergency shutdown, a fire in the zirconium cladding of the fuel bundles could disperse huge amounts of radioactive materials into the atmosphere. Hydrogen explosions could occur, as happened at Fukushima.

Before fuel rods, which have been enriched in uranium-235, are used to create energy, they are barely radioactive. Once that isotope of uranium is bombarded with neutrons in a reactor, however, the fuel rods become highly radioactive. The most potent releases from the Fukushima disaster of 2011 involved the radioactive isotopes iodine-131 and cesium-137. That cesium isotope is damaging to humans and has a half-life of 30 years.

As spent fuel from reactors continues to build up in pools, the reactors' owners have turned to "dry casks" for storage of additional spent fuel. After about five years, when the spent fuel becomes somewhat less radioactive, it can be transferred to large, dry casks, where it is air cooled, a safer method than storage underwater. Storage in pools runs the risk of fires and explosions if the water drains from the pools. Nevertheless, the NRC has allowed reactor owners to keep large amounts of spent fuel in pools and has not required them to transfer greater amounts to dry casks. Older spent fuel is moved to dry casks only when that older fuel is replaced by spent fuel that has just come out of a reactor. Thus, the amount of spent materials remains high in spent-fuel pools in the United States.

On November 12, 2013, the NRC staff sent its commissioners a recommendation titled *Staff Evaluation and Recommendation for Japan Lessons-Learned Tier 3 Issue on Expedited Transfer of Spent Fuel*. It stated, "The staff concludes that the expedited transfer of spent fuel to dry cask storage would provide only a minor or limited safety benefit . . . and that its expected implementation costs would not be warranted. The staff therefore recommends that additional studies and further regulatory analyses of this issue not be pursued, and that this Tier 3 Japan lessons-learned activity be closed" (U.S. NRC 2013).

The NRC staff also determined that "commercial U.S. operating reactor sites typically have greater inventories of spent fuel stored on site than otherwise comparable foreign reactors." The staff analyzed spent-fuel storage in several broad groups. One group consisted of thirty-one GE-type boiling-water reactors, which are similar to the reactors at Fukushima; another group included forty-nine pressurized water reactors, like the reactors at Indian Point. The staff calculated the hazards for each of these broad groups in coming to their conclusions about the expedited transfer of spent fuel and its costs.

The NRC staff should have given special consideration to those reactors where the monetary and human costs of a major release of radioactivity from a spent-fuel pool would be the highest. For these reactors in particular, the cost of implementing expedited fuel transfer to dry casks certainly would be justified.

One NRC staff member filed a nonconcurrence statement on the spent-fuel recommendation of 2013. The reply to him ends with the statement, "The staff followed established processes and guidance [presumably that of the NRC] and provided their findings to the Commission for consideration" (given in U.S. NRC 2013). In my estimation, more emphasis was placed on existing NRC rules than on wise judgments. No indication was given that the NRC also has a responsibility to the legislative and executive branches of the federal government, state governments, and the public.

National Academies Reports on Lessons Learned from Fukushima

The U.S. Congress asked the National Academies of Sciences, Engineering, and Medicine to conduct a review of the Fukushima disaster and the lessons learned from it for the United States. I strongly recommend the National Academies' phase 1 report issued in 2014 and the phase 2 report issued in 2016, both of which focus on spent fuel and reactor security. U.S. committee members traveled to Japan and met with many organizations involved in the Fukushima accident.

The phase 2 report found that Japan received a fortunate break in that young, hot fuel in unit 4 at Fukushima unexpectedly remained covered by water during the first two months of the accident. Unit 4 had been shut down for maintenance, and the entire reactor core and other spent-fuel assemblies were in its pool. Officials in Japan and the United States had feared that spent fuel would become uncovered and release a radioactive plume that could have blanketed a very great swath of northeastern Japan, including Tokyo. Modeling indicated the spent fuel and the active reactor core from unit 4 that was in the pool would have become uncovered and caught fire had there not been a fortuitous, unplanned leakage of water into the pool from other parts of the plant that also were located high up in the same building. Thus, Japan dodged a very dangerous bullet that could have made the Fukushima accident much worse.

The journal *Science* for May 27, 2016, carried the story "Near Miss at Fukushima Is a Warning for U.S., Panel Says" (Stone 2016), which described the results of the phase 2 study and subsequent analysis. The NRC staff made a special analysis of the Peach Bottom reactor in southeastern Pennsylvania, a GE-type boiling-water reactor similar to units 1 to 4 at Fukushima. Frank von Hippel of Princeton, a member of the phase 2 study who led the modeling of a hypothetical spent-fuel fire at Peach Bottom, said, "We're talking about trillion-dollar consequences."

A major spent-fuel fire in the United States could dwarf the Fukushima accident in radioactive release because most U.S. spent-fuel pools are more densely packed. The NRC staff estimated in 2013 that a major fire in the Peach Bottom's spent-fuel pool in southeastern Pennsylvania would displace 3.46 million people from 12,000 square miles (31,000 square kilometers) of land, an area larger than the state of New Jersey. The *Science* article stated, "But von Hippel and others think that the USNRC has grossly underestimated the scale and societal costs of such a fire." The article includes scenarios for spent-fuel fires on four different days at Peach Bottom in 2015, as computed by von Hippel and Michael Schoeppner of Princeton based on the release of radioactive cesium-137. "The contamination from such a fire on U.S. soil 'would be an unprecedented peace-time catastrophe,' the Princeton researchers conclude" (Stone 2016).

The *Science* article also stated, "NRC's first look at the academies report 'did not identify any safety or security issues that would require immediate action' . . . But the benefits of expedited transfer to dry casks are fivefold greater than NRC has calculated, the academies found" (Stone 2016).

U.S. Responses to the Fukushima Disaster

Following the Fukushima accident of 2011, the NRC formed a near-term task force that recommended that owners of nuclear-power plants in the

United States reassess the design bases for seismic events and flooding. A year later the commission formally requested licensees to reevaluate seismic hazards for their reactors using up-to-date information on the propagation of seismic waves and shaking in the eastern and central United States. But similar required work on earthquakes that exceed the "beyond design basis" of reactors had been going on since 2005. Owners were requested to submit a "Seismic Hazard Evaluation and Screening Report" by March 2014, which they did. Entergy and other owners stated in their responses that they did not need to make any repairs or retrofits at Indian Point or other reactor locations.

In December 2014, the NRC staff identified thirty-two reactor sites in the central and eastern United States where the reevaluated hazard exceeded the current seismic design. The NRC said it planned to complete its review with recommendations by the third quarter of 2015. It placed Indian Point in Group 1, "generally those that have the highest re-evaluated hazard relative to the original plant design . . . as well as ground motions in the 1–10 Hz range that are generally higher in absolute magnitude." In May 2014, the NRC stipulated that the owner of Indian Point must submit its Group 1 evaluation by June 30, 2017, and stated that it planned to issue a staff assessment on the reevaluated seismic hazard approximately twelve to eighteen months later, more than seven years after the Fukushima disaster.

In 2014, Senator Ed Markey (D) of Massachusetts, a longtime critic of the NRC, said, "The N.R.C. should be demanding implementation of seismic safety upgrades it called for following the Fukushima meltdowns, not merely more study of nuclear reactors that it knows are clearly at higher risk than was previously believed. What is needed is action to secure at-risk nuclear reactors, not merely more reports." What he describes is unfortunately a typical pattern for the NRC in that although it asked each reactor owner to submit a report, including a "walkdown" of the facility by consultants and analyzed those reports, it did not involve knowledgeable outsiders as critical reviewers.

The U.S. National Academies of Science, Engineering, and Medicine phase 1 report stated, "Nuclear plant licensees and their regulators [in the United States] must actively seek out and act on new information about hazards that have the potential to affect nuclear plant safety" (2014). It went on to recommend efforts by the NRC and the nuclear industry to strengthen capabilities for assessing risks from events that could challenge the design of nuclear-plant structures and lead to a loss of critical safety functions. It concluded that the NRC needs to strengthen the nuclear-safety culture and to improve peer review. It also pointed out that Japanese regulatory agencies were not independent and were subject to "regulatory capture"—that is, the nuclear industry manipulated regulators to put its interests ahead of the public's interest.

In an op-ed article in the *New York Times* on March 24, 2011, von Hippel, who is an expert on arms control, energy, and reactor safety, wrote, "After Fukushima, we can no longer let the nuclear industry resist safety efforts."

The National Academies phase 1 report recommended that the NRC strengthen capabilities for assessing risks from beyond-design-basis events. So far, however, the NRC has allowed reactor owners to respond *voluntarily* to that issue. The report says that risks to reactor-core damage are dominated by beyond-design-basis accidents, including those at Three Mile Island in 1979, Chernobyl in 1986, and Fukushima in 2011.

Seismic Assessments of North Anna

In an extensive article published soon after the Virginia earthquake, investigative reporter Roger Witherspoon (2011) stated that twenty-five of the twenty-seven dry casks at North Anna moved from 0.5 inch to 4.5 inches on their pad during the earthquake and that cracked walls and damage occurred at the base of anchored tanks. Cracking also took place at the base of an important electrical transformer, which caused loss of outside power and immediate station blackout. Fortunately, the plant's on-site generators cooled the reactors and spent-fuel pools until off-site power was restored. One of the two sets of instruments at North Anna for recording strong shaking failed during the 2011 earthquake.

That information is not comforting to me considering what might have happened in a somewhat larger earthquake. Because reactor design and the use of dry casks at Indian Point are similar to those for North Anna, similar effects might be expected at Indian Point in a strong earthquake like that of 2011.

Claims by Entergy and NRC Officials

A spokesperson for Entergy, the owners of Indian Point, claimed those reactors could withstand an earthquake of magnitude 6.0. In a full-page ad in the *New York Times* on March 22, 2011, Entergy stated, "Importantly, Indian Point has been designed to withstand an earthquake 100 times the magnitude of the strongest earthquake that has occurred in the area." This statement shows a complete lack of understanding of earthquake magnitudes. In an email sent to a reporter, Jerry Nappi, a spokesman for Entergy said, "Indian Point is protected not just from the strongest earthquake that has happened, but from the strongest earthquake that is possible." Another official at Indian Point said, "The level of structural malfunction at the Fukushima plant would not happen here." No scientific or technical basis exists to support these contentions.

Prior to the Fukushima disaster, a number of Japanese officials stated that their reactors were completely safe. The possibility of damage to the core of a plant's reactor from an event was considered implausible. As a consequence of this view, planning for such a situation was not treated seriously, leaving Japan unprepared for the scope and extent of the required emergency response in 2011. One Japanese investigator said that the Fukushima accident couldn't be regarded as a natural disaster; it was a profoundly manmade disaster that could and should have been foreseen and prevented. Another said, "You can't adequately prepare for a disaster that you don't admit can ever happen." Also, Japanese officials had not prepared for serious damage to more than one reactor at a site, which occurred at Fukushima One.

In its report of July 17, 2011, the NRC's Near Term Task Force on Fukushima concluded, "A sequence of events like the Fukushima accident is unlikely to occur in the United States and some appropriate mitigation measures have been implemented, reducing the likelihood of core damage and radiological releases. Therefore, continued operation and continued licensing activities do not pose an imminent risk to public health and safety" (given in U.S. NRC 2013). Clearly, the NRC did not learn several crucial lessons from Fukushima.

In 1976, four nuclear engineers, including one from the NRC, resigned, stating that nuclear power was not as safe as their supervisors claimed. They testified, "The cumulative effect of all design defects and deficiencies in the design, construction and operation of nuclear power plants makes a nuclear power plant accident, in our opinion, a certain event." Safety is part of the social contract under which U.S. nuclear plants are allowed to operate. That contract has not been respected.

14

TRAVELS TO EARTHQUAKE COUNTRIES AND A TRIP TO THE EARTH'S MANTLE IN NEWFOUNDLAND

I have had the opportunity to travel extensively throughout the United States and to about sixty other countries since I graduated from college in 1960. Many of my scientific excursions were to interesting and beautiful places. Here I offer highlights of trips of geological and geophysical importance.

The People's Republic of China

In the fall of 1974, I was fortunate to be part of a fourteen-person U.S. seismological delegation invited for a five-week visit to the People's Republic of China. President Richard Nixon had been to China two years earlier in 1972 as a first step in the opening of diplomatic relations between our two countries. He and the Chinese government agreed to exchange visits of a number of delegations in the fields of science, agriculture, and medicine. A Chinese delegation of geophysicists and geologists who worked on earthquake prediction visited the United States early in 1974 and were shown a number of universities and other U.S. sites of interest, with a final visit to Yosemite National Park. In return, the Chinese arranged for our group to see many remote sites of earthquake significance that few, if any, tourists had ever visited. In some of those remote places, Chinese villagers had never seen Westerners or, in fact, any foreigners since Japanese soldiers departed in 1945.

For our final stop, the Chinese arranged for us to visit Guilin, the world-famous site of marvelous eroded mountains of limestone. As our delegation wended through the mountains on the only motorized transport on the Li River, we saw humans slowly pulling boats up the river. Today, many motorized tourist boats now undertake excursions on the river.

During our trip in China, we visited several places in the vicinity of Beijing, including the geophysical and geological institutes, a seismograph factory, the Summer Palace, the Great Wall, the Ming Tombs, and a collective farm that raised Peking ducks. We later had a marvelous meal of about thirteen courses at the restaurant Beijing Duck. In joint presentations with the Chinese in Beijing, many of us talked about the results from work on earthquake prediction and physics. We learned about Chinese work on unusual animal behavior and changes in electrical and magnetic signals before some large earthquakes.

From Beijing, we traveled by train to a small rural city about 300 miles (500 kilometers) southwest of Beijing that recently had been severely damaged by two earthquakes of about magnitude 6.5. Before we left, the Chinese treated us to a concluding banquet in our honor. The next day on a long train trip to Xian in central China, most of us, including me, became sick with intestinal flu. Our interpreter offered to give us traditional Chinese medicine, but when we asked what she took for intestinal sickness, she said tetracycline. During the Cultural Revolution, she and her husband, both intellectuals, had been sent to a farm for reeducation and hard labor.

We flew from Beijing to Kunming in southwestern China, which was the terminus of the Burma Road during World War II and a very active earthquake region. Our plane landed for lunch in Chengdu, the capital of Sichuan Province before flying on to Kunming. From the plane, we had a great view of the eastern side of the massive Tibetan plateau. The Cultural Revolution, nearly over in Beijing in 1974, was still in full blast in Kunming, with many large Chinese character posters on walls all over town. We visited the nearby site of a magnitude 8.0 historic earthquake.

Moscow and Central Asia

In 1971, after a meeting of the International Union of Geodesy and Geophysics in Moscow, I traveled to Samarkand and Bukhara in central Asia as a tourist. I was the only foreigner on an overnight flight. Many of my fellow travelers on the plane ate sausages they had brought with them. When we arrived, I felt that I was indeed very far from home. Samarkand is a beautiful city of old mosques on the ancient Silk Road to China. Bukhara has many magnificently decorated ancient buildings. Both were well worth the trip.

After visiting Samarkand and Bukhara, I went on to Garm, Tajikistan, where I was the guest of the Soviet Academy of Sciences at the headquarters of the Complex Seismological Expedition in Garm, Tajikistan. In Moscow, seismologist Vitaly Khalturin spent many days struggling with Intourist, the Soviet travel agency, to work out my visit to Garm. Intourist did not want to lose any of the "hard currency" I had paid for my trip to Samarkand and Bukhara.

While I was in Garm, I saw work being done on earthquake prediction and earth deformation along the north side of the Hindu Kush and Pamir Mountains to the north of Afghanistan. Measurements on terraces along the north side of those mountains indicated a very high uplift rate of about 0.4 inches (one centimeter) per year. I visited the nearby site of the Khait earthquake of 1949, in which many people had died.

After a long day visiting regional seismograph stations and viewing earthquake records, I was ready to sleep. Nevertheless, a door opened, and the entire team in Garm greeted me with large glasses of vodka. Tanya Rautian, a seismologist and Khalturin's wife, told me to eat, eat, eat from the nearby table after drinking vodka. I was somewhat hung over the next morning, but I went out jogging to wake myself up. I gave a talk on plate tectonics, as I had done in Moscow. My talks were the first time many Russian geophysicists had heard about plate tectonics. I returned to Moscow, Garm, and Samarkand two years later in the fall of 1973 as part of a U.S. delegation on earthquake prediction.

The Dead Sea Plate Boundary and Earthquake Risk in Israel

In 1990, I spent a week in Israel advising its government on earthquake risk and long-term prediction. While I was there, Avi Shapira, my host and the director of the Israeli seismograph network, was kind enough to drive me to many places throughout the country, including the Sea of Galilee (Lake Tiberias), the small Jordan River, and views of the Golan Heights. Other earth scientists took me to Jerusalem, the Dead Sea, and Masada. My views of the Dead Sea rift and Jericho from the heights of Jerusalem were spectacular. Jericho is 846 feet (258 meters) below sea level and may be the oldest continuously occupied city in the world. Faults of the Dead Sea rift system and the African–Arabian plate boundary pass through Jericho. An earthquake may have been the source of "the walls tumbling down" when Joshua's army attacked Jericho.

Earthquakes, Great Faults, and Volcanoes in New Zealand

During a five-month sabbatical in New Zealand starting in November 1991, I took many trips to visit sites that were interesting to me both as a tourist and as a geologist and geophysicist. I especially remember hiking for a very long day up two of the spectacular volcanoes on the north island in National Park. I then visited the thermal hot springs at Rotorua on the North Island, the Wairakei geothermal power station, and Lake Taupo, which was created by a huge volcanic explosion and eruption about 2,000 years ago.

In the South Island, I visited famous fault scarps that were formed in prehistoric earthquakes; Franz Joseph glacier; the Alpine Fault; and Milford Sound, a World Heritage site famous as for its mountainous peaks and biting flies. Many geologists and geophysicists think the Alpine Fault is nearing the end of its cycle of great earthquakes. It has many similarities to California's San Andreas Fault. New Zealand has an enormous number of sheep, a major source of income from export. One resident told me, "New Zealand has 80 million sheep, 3 million of whom think they are people."

A Trip to the Earth's Mantle in Newfoundland

Traveling for ten days in western Newfoundland in 1998, my wife, Kathy, and I spent several days in Gros Morne National Park, where major outcrops of the earth's crust and uppermost oceanic mantle were thrust onto North America about 450 million years ago when a previous ocean closed. We walked to the earth's former oceanic mantle, one of the few places in the world where it is well exposed. We then visited L'Anse Aux Meadows, an authenticated Norse temporary settlement from about the year 1000. The Norse dug pits for their ships, which are now about 3 feet (one meter) higher than their elevation 1,000 years ago, an indication of the rise of the land after the most recent glacier melted and the slow inflow of mantle material into the asthenosphere beneath Newfoundland.

Evolution of Species and Volcanoes in the Galapagos

In May 2000, Kathy and I visited the Galapagos Islands, traveling on an Ecuadorian ship. May was a good month to visit the islands because tourism was not at its peak and many birds were mating, including the albatross, which arrives after flying thousands of miles across the Pacific. Darwin's finches, which figure in his famous book *On the Origin of Species* (1859), differed on various islands, as he stated. Nevertheless, each was a rather drab brown at first glance. The three species of colored boobies, however, were plentiful everywhere and were our favorite birds. Our trip included walks on two active volcanoes in the western Galapagos. This was one of the best trips I have ever taken in terms of scenery, geology, biology, and wildlife. The excellent guides were fully bilingual.

Through all my journeys, both those I have undertaken physically as well as those I have ventured on mentally, I have been fascinated by the challenges and the discoveries that have crossed my path. I am grateful to be part of the scientific community, and I treasure all the people I have met and interacted with along the way.

15

ADVANCES IN LONG-TERM EARTHQUAKE PREDICTION

Future Prospects

Earlier, I discussed the monitoring of seismic gaps and variations of repeat times between large earthquakes for individual fault segments along major plate boundaries in an effort to improve estimates of the timing of future events. A number of new techniques for studying large earthquakes, which I discuss here, have become available in the past 15 years that may bring us additional steps closer to long-term earthquake prediction.

The hypothesis of seismic gaps states that large to great earthquakes are more likely to take place along parts of active plate boundaries where they have not occurred for decades to hundreds of years. Because stresses in the earth accumulate slowly, large shocks at the same place understandably repeat only after an extended period of time. The gap hypothesis, however, provides only qualitative information about the timing of future large shocks. Learning more about precursory phenomena on time scales of about a decade is important, and I explore that topic here.

Table 15.1 lists recent advances using six quite different techniques to identify possible precursory effects on time scales of about 10 years before several well-recorded great to giant earthquakes. Some of these techniques represent new scientific work on my part that is in progress. These techniques may help to advance long-term prediction for at least some segments of active plate boundaries. Instead of emphasizing a "magic bullet" to forecast all large earthquakes, I favor a combination of techniques, each of which may be useful for narrowing the predicted time of at least

TABLE 15.1 Forerunning Phenomena to Great Earthquakes

EARTHQUAKE	METHOD	RESULTS
Tohoku, Japan, 2011 Magnitude 9.0	GPS	Measured before, during, and after 2011 Plate boundary largely locked before
	GPS and slow slip	Decreased compression 1996 to 2011
	Seismic gap	Yes, since at least 1890
	Cumulative moment (summed over time)	Accelerating from 2005 to the main shock in 2011
	b value	Decreased from 2005 to 2011
	Forerunners	Five magnitude 7.0 shocks surrounding epicenter in of coming 2011 shock
	Repeating small shocks	Increased rate from 1996 to the mainshock in 2011
Iquique, Chile, 2014 Magnitude 8.1	GPS	Measured before, during, and afterward Plate boundary largely locked before
	GPS and slow slip	Coastal stations accelerated toward trench eight months beforehand
	Seismic gap	Yes, since giant earthquake in 1877
	Cumulative moment	Accelerating from 2007 to the mainshock in 2014
	Foreshocks	Abundant at shallow depths
Maule, Chile, 2010 Magnitude 8.8	GPS	Measured before, during, and after Plate boundary mostly locked beforehand
	Seismic gap	Yes, from great shocks in 1835 and 1906 to 2010
	Cumulative moment	No clear acceleration but forerunning earthquakes at ends of rupture zone
Illapel, Chile, 2015 Magnitude 8.2	GPS	Measured before, during, and afterward Plate boundary largely locked before
	Seismic gap	Yes, since great earthquake of 1943
	Cumulative moment	No clear acceleration but forerunning activity found throughout rupture zone
San Francisco, 1906 Magnitude 7.7	Cumulative moment	Accelerating from 1883 to mainshock of 1906

some coming large shocks. The identification of slow-slip events may be a "magic bullet," but additional research is needed to determine whether it is useful for prediction.

Value of GPS Data

During the past 30 years, GPS data from satellites have revolutionized the study of slip during large earthquakes and of stress buildup. Local seismic and GPS stations were operating in the rupture zones of the Maule, Tohoku, Iquique, and Illapel earthquakes in 2010, 2011, 2014, and 2015 before, during, and after the events. Each of the four events occurred in long-existing seismic gaps. GPS observations were essential in demonstrating that the four coming rupture zones were largely locked—that is, not slipping—in the preceding period of slow stress buildup (figure 15.1).

In 2015, John Loveless and Brendan Meade used GPS data to calculate the percentage of locking (coupling) along the main plate boundaries of

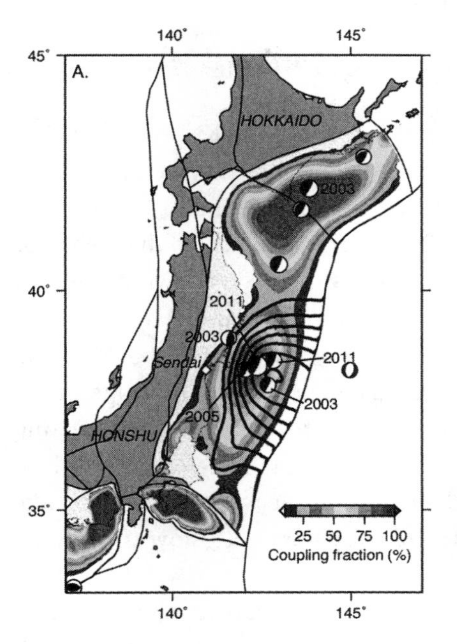

FIGURE 15.1 Strong seismic coupling along several parts of plate boundaries off Japan as determined from GPS observations. See plate 28.

Source: Loveless and Meade 2015.

Japan. They found strong coupling between 37°N and 39°N (figure 15.1) in the zone that broke in 2011. They also calculated strong coupling between 40°N and 43°N, the site of great shocks in 1968 and 2003, as well as along the Nankai subduction zone farther south, which most recently ruptured in great earthquakes in 1944 and 1946 (figure 10.3). They found poorer coupling in the region off Honshu between 39°N and 40°N.

It is essential to know which segments of plate boundaries are locked and which ones, like the intersection of the Carnegie Ridge with the South American subduction zone (figure 7.4), appear to be slowly slipping over the long term. The few existing GPS observations indicate that the Carnegie Ridge segment does not have the potential to be the site of great future shocks. In contrast, seismic gaps that are fully locked still remain to the north and south of the northern Chilean rupture zone of 2014 (figure 7.5), so they could well be sites of future earthquakes of magnitude 8.0 to 8.5.

Future studies of seismic gaps aimed at long-term prediction need to have both local seismic and GPS observations. As time progresses, longer and more accurate GPS data sets likely will become available; GPS instruments are likely to be installed along more plate boundaries.

Slow-Slip Events Before Large Earthquakes

Using GPS data, Andreas Mavrommatis, Paul Segal, and Kaj Johnson (2014) found that the rate of compression of the upper plate and stress buildup slowed from 1996 until the giant Tohoku earthquake in 2011 (figure 10.8). Yusuke Yokota and Kazuki Koketsu (2015) placed the start of that behavior around 2001. Japanese geophysicists found a similar slowdown using data from offshore stations (figure 10.7). Mavrommatis and his colleagues deduced that slow slip along the plate boundary at depths of about 12 to 30 miles (20 to 50 kilometers) led to decreased compression of the upper plate. In 2015, he and other colleagues also found that the frequency of repeating small earthquakes increased in the coming rupture zone of the shock in 2011. Both observations as well as the results of the work done by Yusuke Yokota and Kazuki Koketsu in 2015 are consistent with slow slip having commenced at depth.

In the absence of contrasting data, stresses at the surface previously had been assumed to build up uniformly. We now know that slow slip, which occurs without shaking, appears to be a significant component of total fault slip. Slow-slip events and changes in the rates of buildup of stresses along plate boundaries are exciting areas of recent research that may provide additional precursory information for time scales of years to decades. Nevertheless, we need to know if slow-slip events of those durations happen only or mainly during the last stage of stress accumulation before great earthquakes or if they occur during other parts of the period of slow stress buildup.

Accelerated Seismic-Moment Release Prior to Some Great Shocks

Figure 15.2 depicts cumulative seismic-moment release in the decades preceding the great earthquake near Iquique in northern Chile in 2014. *Seismic*

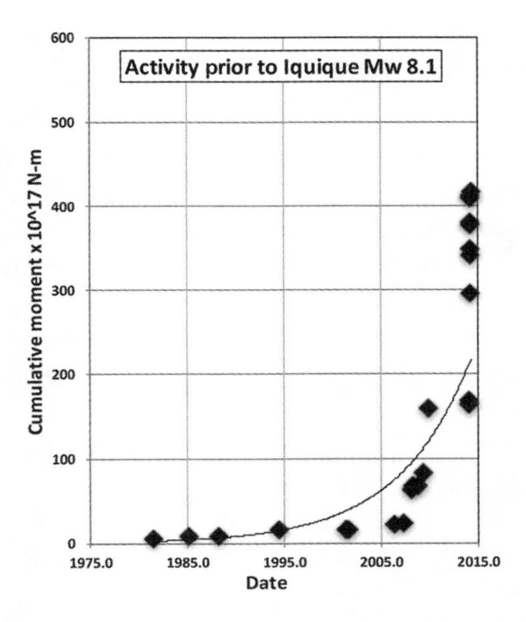

FIGURE 15.2 Cumulative seismic moment release as a function of time before the Chilean earthquake of 2014. Thin black line is the best-fitting exponential increase in summed moment with time.

Source: Unpublished figure by the author, 2016.

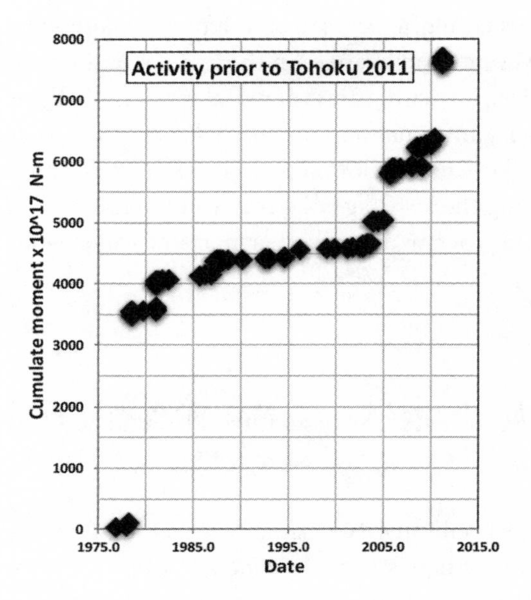

FIGURE 15.3 Cumulative seismic moment release from 1976 until the Tohoku earthquake of March 2011.

Source: Unpublished figure by the author, 2016.

moment is calculated from the very long-period seismic waves generated by earthquakes. If moment release occurred uniformly, its cumulative sum with time would have defined a straight line increasing upward, not the exponential increase observed. A great increase in moment release was most dramatic in the preceding one-half month to three months before the great shock. It could have been the basis for a short-term warning. As early as 2009, such increases might have been used to declare an intermediate-term prediction.

Figure 15.3 shows the cumulative release of seismic moment from 1976 until just before the giant Japanese earthquake of 2011. Although a large increase in seismic moment occurred during the magnitude 7.6 shock in 1978, little additional moment was accumulated until four earthquakes of magnitude 7.0 to 7.2 took place between May 2003 and August 2005. A substantial acceleration in seismic-moment release occurred between May 2003 and 2010. This finding might have been the basis for an intermediate-term prediction. With that in mind, the substantial increase in moment release in the days before the giant shock might have been the basis for an enhanced earthquake watch, if not for a short-term prediction.

In 1990, Lamont graduate student Steven Jaumé and I reported that moderate-size shocks were abundant and felt in a broad region surrounding the coming rupture zone of the San Francisco earthquake in 1906, especially between 1883 and 1906 (Sykes and Jaumé 1990) (figure 15.4). The rate of occurrence of events dropped dramatically after 1906, as shown for a period of the same length from 1920 to 1954. Activity picked up again from 1955 until just before the Loma Prieta earthquake of 1989. The latter activity, however, was concentrated in a much smaller region to the southeast of San Francisco around the coming earthquake of magnitude 6.9 in 1989 than the activity before the much larger event in 1906. A similar pattern was found prior to the magnitude 6.8 shock along the Hayward fault in 1868.

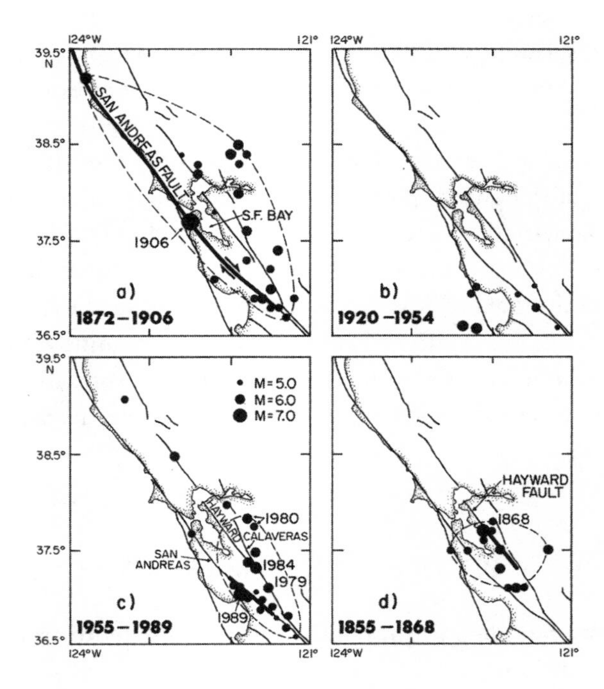

FIGURE 15.4 Distribution of earthquakes of magnitude 5.0 and larger in the San Francisco Bay region for four time periods. Major faults are shown as thin solid lines. Thick solid lines indicate fault displacements (slip) at the surface in the large shocks of 1868, 1906, and 1989.

Source: Modified from Sykes and Jaumé 1990.

Jaumé and I also reported that cumulative seismic-moment release grew exponentially for about 10 years prior to the large California events of 1868, 1906, and 1989. We stated that it might be possible to utilize such increases to predict similar large earthquakes on time scales of a few years to a decade.

In 1999, my Lamont colleagues Bruce Shaw and Christopher Scholz and I cited the increased numbers of moderate-size earthquakes before the three large shocks in the San Francisco Bay region as examples of what has been called *critical organized behavior* on a regional scale (Sykes, Shaw, and Scholz 1999). Such behavior is thought to occur before various major natural events as well as prior to major sand avalanches on a sand pile in a laboratory. In the latter case, moderate-size sand avalanches occurred more often in a large region prior to a major avalanche.

We reported that the stresses in the areas surrounding the rupture zones of individual large earthquakes were reduced below a critical state of stress after previous large events occurred and remained so for long periods. As stresses were slowly reestablished by plate tectonic loading, each of the surrounding regions approached a critical state prior to failure. The presence of such a critical state can be regarded as a long-term precursor.

It is significant that many and perhaps all of the forerunning seismic activity shown in figure 15.4 occurred on nearby but different faults than the ones that actually ruptured in 1868, 1906, and 1989. Although many investigators have claimed they have found little or no forerunning seismic activity on faults that had just ruptured in large shocks, the examples from the San Francisco area indicate that searches for forerunning activity should be broader and made in regions that surround a suspected future large earthquake, such as along a specific fault—for example, the San Andreas.

Precursory Changes in *b* Values

Seismologists call one measure of the relative number of small to larger earthquakes the *b value*. Calculated from the numbers of shocks of various magnitudes, in most instances the *b* value is close to 0.9 to 1.0. In the cases discussed here, the number of larger shocks increases in the period leading up to a great earthquake, whereas the number of smaller events remains about the same. In rock-mechanics experiments, significantly lower *b* values have been associated with high stresses. In 2014, Bernd Schurr of

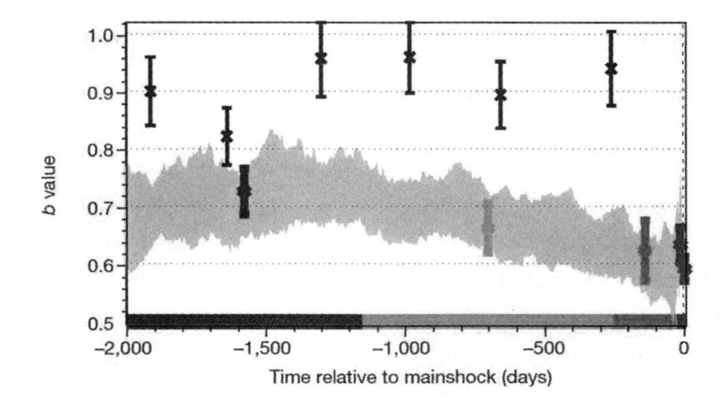

FIGURE 15.5 Decrease in *b* values in coming rupture zone 5.5 years before Iquique mainshock of 2014. Black bars indicate higher, normal values in adjacent areas outside of that rupture zone. See plate 29.

Source: After Schurr et al. 2014.

the German Centre for Geosciences in Potsdam and his colleagues found that *b* values were anomalously low for 5.5 years in the coming rupture zone of the great Chilean earthquake of 2014. They found larger—that is, normal—*b* values outside that zone (figure 15.5), which suggested that the coming rupture zone in Chile was the site of high stress.

In 2017, Anne Socquet of the Université Grenoble Alpes and her colleagues reported that a group of coastal GPS stations accelerated toward the trench in the eight months prior to the earthquake of 2014, much like the acceleration observed before the Japanese shock of 2011 depicted in figure 10.8. Hence, precursory forerunning changes in *b* value, accelerating slow slip, seismic-moment release, and moderate-size shocks occurred prior to the mainshock of 2014.

In 2012, K. Z. Nanjo and three other Japanese geophysicists showed that *b* values decreased from a normal value of about 1.1 in the 20 to 5 years before the giant Tohoku earthquake of 2011. They found that the *b* value then decreased further in the five years before 2011 to a very low value of 0.45 within the coming rupture zone. They also reported a steady decrease in *b* value for 30 years prior to the giant Sumatran earthquake of 2004. Their results indicate that high stresses were present near those two coming rupture zones. Thus, *b* values can be interpreted as stress meters and long-term precursors.

Recognizing Foreshocks Before Great Earthquakes: Role of Asperities

If a fault segment were of uniform strength throughout and did not consist of strong zones of fault contact called *asperities* surrounded by weaker zones, smaller forerunners to large earthquakes probably would not occur. The distribution of weak and strong zones of various sizes on a fault, however, can lead to precursory changes before large shocks, which makes long-term prediction possible, at least in theory.

Foreshocks usually have been recognized only with hindsight after large earthquakes have occurred. They rarely have been distinguished ahead of time from the many other small to moderate-size ongoing events in a region. The "trick" before a great shock is to identify and map large asperities—that is, strong zones of fault contact, which are locked and remain relatively quiet until they are loaded and triggered either by foreshocks of moderate-size in adjacent weaker regions or by slow seismic slip. Because the largest strongly coupled parts of a plate boundary usually are seismically mostly quiet before a great earthquake, a reasonable question is, Can large asperities be identified beforehand?

Foreshocks preceding the earthquake in Iquique, northern Chile, in 2014 were relatively easy to identify ahead of time because (1) they took place in a seismic gap that had not ruptured in a great earthquake since 1877; (2) they occurred far from the aftershocks of the great earthquakes of 2001 off southern Peru and 2007 off Chile; and (3) they were numerous. Those foreshocks occurred at shallower depth than the strong region closer to the coast that was the site of anomalously low *b* values.

The unusual clustering (figure 15.6) of four magnitude 7.0 to 7.6 shocks by August 2005 within about 30 miles (50 kilometers) of the epicenter of the coming Tohoku earthquake in 2011 also can be interpreted as a long- to intermediate-term precursor. That region was unusual in having several very large asperities (strong, well-coupled regions) that ruptured in events larger than magnitude 7.0 prior to the mainshock in 2011 (figure 15.6).

Officials in Japan might have used that cluster of large earthquakes, small *b* values, and precursory slip to declare an "earthquake watch," similar to the lowest levels of warnings for tornadoes, hurricanes, and winter storms in the United States. By not issuing any warnings, Japan paid a huge human price of 18,000 deaths and asset losses of more than U.S.$235 billion in 2011.

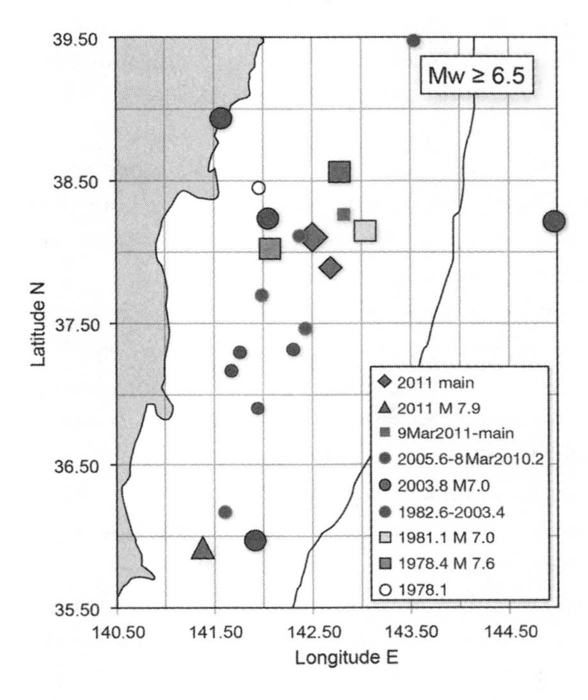

FIGURE 15.6 Earthquakes of magnitude 6.5 and larger from 1976 through the occurrence of the giant earthquake on March 11, 2011, and its largest aftershock of magnitude 7.9. Larger symbols denote earthquakes of magnitude 7.0 and greater. See plate 30.

Source: Unpublished figure by the author, 2016.

The declaration of an earthquake watch six years earlier for an earthquake of about magnitude 8.0 to 8.5 might have engendered greater preparedness, including possibly moving the standby generators at Fukushima to higher ground and providing for more reliable off-site power.

Use of Repeat Times of Large Historic and Prehistoric Earthquakes for Better Long-Term Forecasts

The field of paleoseismology, in its infancy in 1979, has been critical in providing approximate dates, sizes, and amounts of slip in large prehistoric earthquakes. For most Californian fault segments, only one or no large events are known from historic records. Excavations across some fault segments in California since 1979 led to the identification and approximate dating of as many as ten large prehistoric shocks, as described in chapter 8.

In 2006, William Menke of Lamont and I examined repeat times of large earthquakes along fault segments of active plate boundaries in Alaska, Japan, California, Cascadia, and Turkey. When a long series of historic shocks is known, as in Japan, we found that great events occur nearly periodically. Exact historical dates like those provide information on the natural variability of repeat times of large earthquakes along a given fault segment. For most of the segments we examined, however, the sets of repeat times are dominated by paleoseismic determinations whose dates are only approximate because their carbon-14 ages are uncertain. Menke and I devised a procedure to use the published uncertainties in dating to calculate natural, uncontaminated variations in repeat times—that is, to remove the effect of uncertainties in carbon-14 dating (Sykes and Menke 2006).

We calculated the intrinsic repeat time and a measure of its uncertainty, called the *coefficient of variation* (CV), for each of the fifteen fault segments in figure 15.7. CV is the uncertainty (standard deviation) of repeat times for a given fault segment divided by its average repeat time. Because average repeat times varied from as short as 23 years for Parkfield, California, to as long as 700 years for the northeastern part of the rupture zone in Alaska in 1964, we used CV rather than the standard deviation to compare our results among various fault segments.

Many of our determinations of CV in figure 15.7 are quite small—0.08 to 0.26. Thus, large earthquakes along those segments have occurred quasi-periodically in time. That near periodicity gives us hope for long-term prediction. The larger CV of 0.37 for historic shocks at Parkfield indicates that shocks do not occur there at as regular intervals as the other fault segments shown in figure 15.7. This accounts in part for the USGS's failure to predict the Parkfield earthquake of 2004.

The most recent large shock along the Hayward Fault in the San Francisco Bay Area occurred in 1868. Our calculations indicate that this fault is well advanced in its cycle of stress accumulation leading to the next large earthquake. The calculated probability of a repeat of a large earthquake along it by 2029 is about 68 percent. The Hayward Fault crosses densely populated and industrialized areas. Nevertheless, one of the societal problems for governmental officials and the public is taking such long-term predictions seriously.

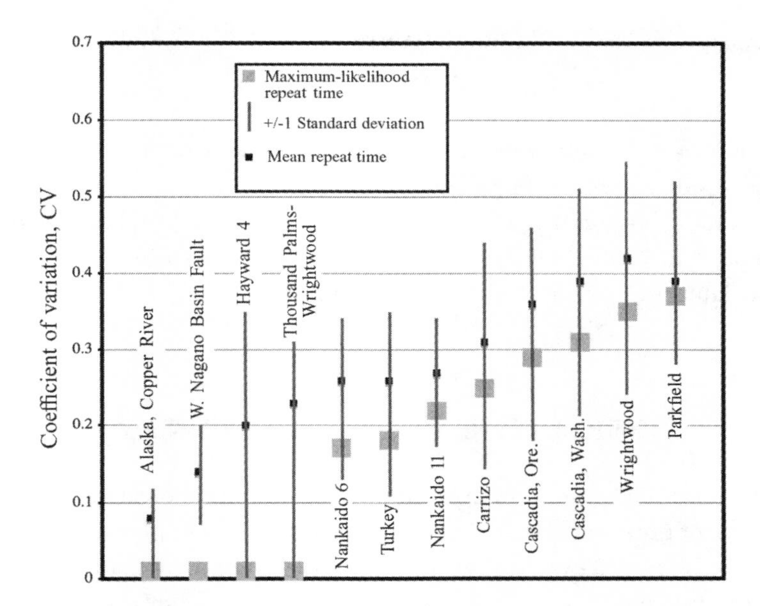

FIGURE 15.7 Coefficient of variation, CV, a measure of the uncertainty in individual time intervals between large shocks that have ruptured various segments of active plate boundaries. See plate 31.

Source: Sykes and Menke 2006.

The segment of the North Anatolian Fault near Gerede, Turkey, has a CV of 0.26 and an average repeat time of 310 years. Dates of past earthquakes at Gerede extend back more than 2,000 years. The fault most recently ruptured in a large event of magnitude 7.6 in 1944, so it is not likely to be the site of another large shock for a few centuries. The northeastern end of the fault zone that ruptured in Alaska in 1964 has an average repeat time of about 700 years and an uncertainty (standard deviation) of 55 years. This fault, too, is unlikely to be the site of another giant shock for centuries. Estimates like these two represent useful predictions that large shocks will *not occur* in given time intervals.

We now recognize that several segments of transform faults along the fast-spreading East Pacific Rise rerupture every 5 to 20 years in shocks of about magnitude 5.0 to 6.0. Because these repeating earthquakes with much shorter time scales and smaller size occur more often than large events in California, they represent an excellent laboratory for learning about long-term prediction.

Increases and Decreases in Stresses Along Faults from Nearby Earthquakes

One advance in computing in the past few decades can be used to determine whether stresses along a fault have either increased or decreased when one or more large earthquakes occur in its vicinity. In 1997, Lamont graduate student Jishu Deng and I computed changes in stress over time for many faults in southern California since 1812.

Prospects for Long-Term Prediction: A Summary

I have described several forerunning effects that have occurred before a number of great and giant earthquakes. Many of those phenomena have been well recorded by many seismic and GPS stations—a situation that did not exist 40 years ago when discussions about earthquake prediction focused largely on data from single observation points. It is clear now that data from GPS and local seismic stations are needed to move beyond identification of seismic gaps to ascertain if long and intermediate-term precursors can be detected and then used for socially useful forecasts.

Slow-slip events, which were virtually unknown more than 25 years ago, are now being widely reported. Some of them have preceded large earthquakes by more than a decade. They represent one hope for long-term prediction. We need to understand, however, if they occur mainly before large shocks or often at other times in the earthquake cycle.

The sizes and spacing of strong regions, or asperities, clearly vary among segments of plate boundaries. The largest ones that ruptured in the decades prior to the earthquake in Japan in 2011 were tens of kilometers in size and of at least magnitude 7.0. Such large forerunners have not preceded some other great earthquakes.

The cumulative release of seismic moment in the rupture zone from 1890 until just before that giant Japanese earthquake in 2011 was not large enough to relieve stresses that had built up slowly during that 121-year interval. By 2000, GPS data indicated quite clearly that the plate boundary in that rupture zone was mostly locked. Hence, a great or giant earthquake was needed to relieve pent-up stresses. At least a decade earlier, the plate boundary that broke in 2011 should have been regarded as a major seismic gap that had not ruptured in a great earthquake since the beginning of instrumental seismic recordings in the 1890s.

Many of my colleagues in geophysics in the United States seem afraid to utter the "*p* word"—*prediction*. The stakes are too great, however, in many places to ignore working on long-term prediction. We need to cast predictions or forecasts in probabilistic formats. We may be able to provide long-term predictions for some but not all earthquakes. We will make mistakes, but we will be able to learn from them. Because the word *prediction* is so loaded, it may be better to call long and intermediate-term predictions either *earthquake watches* or *earthquake alarms*. The times when alarms expire need to be stated clearly.

The data I have described here lead me to be hopeful that we are gradually getting closer to long- and perhaps intermediate-term predictions for at least some great earthquakes along well-monitored subduction zones. Nevertheless, we still have far to go in gaining an understanding of the physics of large earthquakes and in deciding if long- and intermediate-term predictions or earthquake watches are achievable for at least some places. With the placement of monitoring instruments along many seismic gaps, it should be possible in about 30 years to "catch" several more great shocks in very local networks of seismic, GPS, and other geophysical phenomena. This is an exciting time for work on earthquakes.

ACKNOWLEDGMENTS

I thank my wife, Kathleen M. Sykes, for her careful editing and advice. I thank Dan Davis, Kevin Krajick, Stuart Nishenko, and Seth Stein for reading the manuscript. I worked with Davis and Nishenko for many years on various aspects of plate tectonics and large earthquakes as well as with John Armbruster, Göran Ekström, Bryan Isacks, Klaus Jacob, John Kelleher, Won-Young Kim, William McCann, William Menke, Jack Oliver, Walter Pitman, Marc Sbar, Leonardo Seeber, Chris Scholz, and Bruce Shaw.

Ron Doel, now of Oregon State University, recorded seven sessions he had with me in 1996 and 1997. This oral history includes my childhood, education, the development of plate tectonics, and my work on nuclear-test verification. Each session is available through either the Oral History Research Office of Columbia University or the American Institute of Physics. The American Institute of Physics transcribed the tapes, but I was not given the opportunity to correct spelling and other minor mistakes before the transcriptions were placed on the institute's website.

In 1999, Lonnie Lippsett helped me edit some of the material I had written on my personal life for a publication on the fiftieth anniversary of the founding of the Lamont-Doherty Earth Observatory.

My mother put together a baby book of me, which contains my footprints a few days after birth and very early photos. I will be donating it and other documents in my personal collection to the Rare Book and Manuscript Library at Columbia University.

GLOSSARY

Asthenosphere: Layer of low long-term strength in the earth.

Capable fault: Definition used by the U.S. Nuclear Regulatory Commission for an active fault.

Continental drift: Movement of continents through geological time.

Core of the earth: Deepest part of the earth; mostly iron; beneath mantle.

Core of a reactor: See *reactor core.*

Crust: Outermost layer of the earth, of different composition than the mantle.

Comprehensive Nuclear Test Ban Treaty: An international ban on the testing of all nuclear explosions.

Design basis: Range of conditions and events taken into account in the design of a nuclear-power reactor.

Dilatation: Inner movement at the source of an earthquake.

Dry-cask storage: Air-cooled device for storing spent nuclear fuel.

Earthquake hazard: Levels of shaking in an earthquake for a given period of time.

Earthquake risk: Earthquake hazard multiplied by assets (people and structures) and by their vulnerability.

Epicenter: Location of a seismic event in latitude and longitude.

Fault plane: Surface within earth along which displacement takes place.

Fracture zone: Region close to a fault where rocks have been broken in past earthquakes.

Hypocenter: Location of a seismic event in three dimensions.

Isotopes: Varieties (flavors) of an element with different numbers of neutrons but the same number of protons in its atomic nucleus.

Lithosphere: Layer of high long-term strength in the earth.

Magma: Hot liquid rock.

Magnitude: A measure of the size of a seismic event.

Mantle: The part of the earth between the crust and the core.

Mohorovicic (Moho or M) discontinuity: Boundary between the crust and the uppermost mantle of the earth.

Meltdown: Large-scale melting of fuel rods in the core of a nuclear reactor.

Moment magnitude (Mw): Very long-period seismic magnitude.

Microearthquakes: Earthquakes smaller than about magnitude 3.0.

Normal fault: An inclined fault along which displacement is only in the dip direction.

Nuclear-arms control: Limits or restraints on nuclear weapons and/or their delivery vehicles that are mutually agreed upon between states.

Nuclear Regulatory Commission (NRC): U.S. agency charged with formulating policy and regulations, issuing licenses, and overseeing safety for U.S. nuclear reactors.

Reactor core: Central part of a reactor containing fuel assemblies and control rods; where fission occurs.

Reprocessing: Chemical separation of components of spent fuel to obtain plutonium and remaining uranium.

Richter, Charles: Seismologist who devised the earthquake magnitude scale.

Seismicity: Descriptions of locations and sizes of earthquakes and their relationship to geological features.

Seismology: Study of earthquakes and earth structure.

Spent-fuel pool: Structure containing spent-fuel assemblies underwater.

Spent nuclear fuel: Fuel that can no longer be used effectively for the generation of electricity.

Strike-slip fault: A fault along which one side slips horizontally with respect to the other side.

Subduction: The underthrusting of one plate beneath another.

Tectonics: Discipline of geology involving processes that control structure, deformation, and properties of the earth, moons, and other planets.

Transform fault: A strike-slip fault that ends—that is, transforms into either a Mid-Oceanic Ridge or a subduction zone.

Tsunami: In oceans or bays, waves generated by earthquakes and occasionally by volcanic eruptions and landslides.

World-Wide standardized seismograph network (WWSSN): A global network of 125 stations installed in 1962 and 1963, each with identical Press-Ewing seismometers that measured long-period waves and Benioff seismographs for detecting short-period waves.

REFERENCES

Atomic Safety and Licensing Appeal Board, U.S. Nuclear Regulatory Commission (NRC). 1977. *Decision on Geologic and Seismological Issues for Indian Point.* ALAB-436, NRC doc. ML1108700042. Washington, D.C.: NRC, October 12.

Daly, Reginald A. 1940. *Strength and Structure of the Earth.* New York: Macmillan.

Davis, J. F. 1974. "Statement Regarding Licensing of Indian Point Reactor #3 and Discussion of the Final Safety Analysis Report Sections 2.7 (Geology) and 2.8 (Seismology)." New York State Museum and Science Service, Albany, April 19.

Delouis, B., J-M. Nocquet, and M. Vallée. 2010. "Slip Distribution of the February 27, 2010, Mw = 8.8 Maule Earthquake, Central Chile, from Static and High-Rate GPS, InSAR, and Broadband Teleseismic Data" *Geophysical Research Letters* 37:L17305.

Deng, J., and L. R. Sykes. 1997. Evolution of Stress Field in Southern California and Triggering of Moderate-Size Earthquakes: A 200-Year Perspective." *Journal of Geophysical Research* 102:9859–86.

Eberhart-Phillips, D., P. J. Haeussler, J. T. Freymueller, A. D., Frankel, C. M. Rubin, P. Craw, N. A. Ratchkovski, et al. 2003. "The 2002 Denali Fault Earthquake, Alaska: A Large Magnitude, Slip-Partitioned Event." *Science* 300:1113–18.

Entergy. 2011. Full-page ad. *New York Times*, March 22.

Farrar, Chairman M. C., Atomic Safety and Licensing Appeal Board, U.S. Nuclear Regulatory Commission (NRC). 1979. *In the Matter of Consolidated Edison Company of New York, Inc., and Power Authority of the State of New York, Memorandum* [dissenting opinion], *Indian Point Stations, Units 1, 2, and 3.* August 3.

Frankel, H. R. 2012. *The Continental Drift Controversy.* 4 vols. Cambridge: Cambridge University Press. (These volumes contain an excellent and extensive summary of seafloor spreading, transform faulting, confirmation of continental drift, and plate tectonics. Frankel, an historian of science, interviewed many of those involved in discoveries in these areas and includes unpublished letters and drafts of manuscripts.)

Hayes, G. P., M. W. Herman, W. D. Barhart, K. P. Furlong, S. Riquelme, H. M. Benz, E. Bergman, S. Barientos, P. S. Earle, and S. Samsonov. 2014. "Continuing Megathrust Earthquake Potential in Chile After the 2014 Iquique Earthquake." *Nature* 512:295–98.

Isacks, B. L., J. Oliver, and L. R. Sykes. 1968. "Seismology and the New Global Tectonics." *Journal of Geophysical Research* 73:5855–99.

James, T. S., J. F. Cassidy, G. C. Rogers, and P. J. Haeussler. 2015. "Introduction to the Special Issue on the 2012 Haida Gwaii and 2013 Craig Earthquakes at the Pacific–North America Plate Boundary (British Columbia and Alaska)." *Bulletin of the Seismological Society of America* 105, no. 2B: 1053–57.

J. L. Witt Associates. 2003. *Review of Emergency Preparedness of Areas Adjacent to Indian Point and Millstone*. Prepared for Power Authority of State of New York. Albany: State of New York.

Katsumata, M., and L. R. Sykes. 1969. "Seismicity and Tectonics of the Western Pacific: Izu-Mariana-Caroline and Ryukyu-Taiwan Regions." *Journal of Geophysical Research* 74:5923–48.

Kelleher, J., L. Sykes, and J. Oliver. 1973. "Possible Criteria for Predicting Earthquake Locations and Their Application to Major Plate Boundaries of the Pacific and the Caribbean." *Journal of Geophysical Research* 78:2547–85.

Kim, W. Y., L. R. Sykes, J. H. Armitage, J. K. Xie, K. H. Jacob, P. G. Richards, M. West, F. Waldhauser, J. Armbruster, L. Seeber, W. X. Du, and A. Lerner-Lam. 2001. "Seismic Waves Generated by Aircraft Impacts and Building Collapses at World Trade Center, New York City." *EOS Transactions of the American Geophysical Union* 47:565–71.

Koper, K. D., A. R. Hutko, T. Lay, C. J. Ammon, and H. Kanamori. 2011. "Frequency-Dependent Rupture Process of the 2011 Mw 9.0 Tohoku Earthquake: Comparison of Short-Period P Wave Backprojection Images and Broadband Seismic Rupture Models." *Earth Planets Space* 63:599–602.

Krajick, K. 2016. "A Morning That Shook the World: The Seismology of 9/11." Earth Institute, Columbia University, blog, September 6. http://blogs.ei.columbia.edu/2016/09/06/a-morning-that-shook-the-world.

Kubo, H., and Y. Kakehi. 2013. "Source Process of the 2011 Tohoku Earthquake Estimated from the Joint Inversion of Teleseismic Body Waves and Geodetic Data Including Seafloor Observation Data: Source Model with Enhanced Reliability by Using Objectively Determined Inversion Settings." *Bulletin of the Seismological Society of America* 103:1195–220.

Lawson, Chairman A. C. [1908] 1969. *The California Earthquake of April 18, 1906: Report of the State Earthquake Investigation Commission*. Publication no. 87, vol. 1. Washington, D.C.: Carnegie Institution of Washington.

Lay, T., H. Yue, E. E. Brodsky, and C. An. 2014. "The April 2014 Iquique, Chile, Mw 8.1 Earthquake Rupture Sequence." *Geophysical Research Letters* 41:318–25.

Le Pichon, X. 1968. "Sea-Floor Spreading and Continental Drift." *Journal of Geophysical Research* 73:3661–97.

Lienkaemper, J. J., P. L. Williams, and T. P. Guilderson. 2010. "Evidence for a 12th Large Earthquake on the Southern Hayward Fault in the Past 1900 Years." *Bulletin of the Seismological Society of America* 100:2024–34.

Lochbaum, D., E. Lyman, S.Q. Stranahan, and the Union of Concerned Scientists. 2014. *Fukushima: The Story of a Nuclear Disaster*. New York: New Press.

Lomnitz, C. 1970. "Major Earthquakes and Tsunamis in Chile During the Period 1535 to 1955." *Geologischen Rundschau* 59:938–60.

Loveless, J. P., and B. J. Meade. 2015. "Kinematic Barrier Constraints on the Magnitudes of Additional Great Earthquakes off the East Coast of Japan." *Seismological Research Letters* 86:202–9.

Mavrommatis, A. P., P. Segall, and K. M. Johnson. 2014. "A Decade Scale Deformation Transient Prior to the 2011 Mw 9.0 Tohoku-Oki Earthquake." *Geophysical Research Letters* 41:4486–94.

Mavrommatis, A. P., P. Segall, N. Uchida, and K. M. Johnson. 2015. "Long-Term Acceleration of Aseismic Slip Preceding the Mw 9 Tohoku-Oki Earthquake, Constraints from Repeating Earthquakes." *Geophysical Research Letters* 42:9717–25.

Mazzotti, S., and J. Townsend. 2010. "State of Stress in Central and Eastern North American Seismic Zones." *Lithosphere* 2:76–83.

McCann, W. R., S. Nishenko, L. R. Sykes, and J. Krause. 1979. "Seismic Gaps and Plate Tectonics: Seismic Potential for Major Earthquakes." *Pure and Applied Geophysics* 117:1082–147. Also published as U.S. Geological Survey Open File Report 78-943 (1978).

McEnaney, D. D. 2001. *The Descendent of Jacob Sykes III*. N.p.: self-published. (Jacob Sykes III [1829–1910] was my and McEnaney's great-great-grandfather.)

McKenzie, D., and R. L. Parker. 1967. "The North Pacific: An Example of Tectonics on a Sphere." *Nature* 216:1276–80.

Meltzer, A. J., K. Sieh, H.-W. Chiang, C.-C. Wu, L. L. H. Tsang, C.-C. Shen, E. M. Hill, B. W. Suwargadi, D. H. Natawidjaja, B. Philibosian, and R. W. Briggs. 2015. "Time-Varying Interseismic Strain Rates and Similar Seismic Ruptures on the Nias-Simeulue Patch of the Sunda Megathrust." *Quaternary Science Reviews* 122:258–81.

Molnar, P., and L. R. Sykes. 1969. "Tectonics of the Caribbean and Middle America Regions from Focal Mechanisms and Seismicity." *Bulletin of the Geological Society of America* 80:1639–84.

Moreno, M., M. Rosenau, and O. Oncken. 2010. "2010 Maule Earthquake Slip Correlates with Pre-seismic Locking of Andean Subduction Zone." *Nature* 467:198–202.

Morgan, W. J. 1968. "Rises, Trenches, Great Faults, and Crustal Blocks." *Journal of Geophysical Research* 73:1959–82.

Munk, W. H., and G. J. F. MacDonald. 1960. *The Rotation of the Earth: A Geophysical Discussion*. Cambridge: Cambridge University Press.

Nanjo, K. Z., H. Hirata, K. Obara, and K. Kasahara. 2012. "Decade-Scale Decrease in b Value Prior to the M9-Class 2011 Tohoku and 2004 Sumatra Quakes." *Geophysical Research Letters* 39:20304–312.

Nishenko, S. P. 1985. "Seismic Potential for Large and Great Interplate Earthquakes Along the Chilean and Southern Peruvian Margins of South America: A Quantitative Reappraisal." *Journal of Geophysical Research* 90:3589–615.

Nishenko, S. P., and G. A. Bollinger. 1990. "Forecasting Damaging Earthquakes in the Central and Eastern United States." *Science* 249:1412–16.

Oliver, J., and B. Isacks. 1967. "Deep Earthquake Zones, Anomalous Structures in the Upper Mantle, and the Lithosphere." *Journal of Geophysical Research* 72:4259–75.

Pacheco, J. F., and L. R. Sykes. 1992. "Seismic Moment Catalog of Large, Shallow Earthquakes, 1900–1989." *Bulletin of the Seismological Society of America* 82:1306–49.

Pitman, W. C., III, and J. R. Heirtzler. 1966. "Magnetic Anomalies Over the Pacific-Antarctic Ridge." *Science* 154:1164–71.

Plafker, G. 1964. "Tectonic Deformation Associated with the 1964 Alaska Earthquake." *Science* 148:1675–87.

Richter, Charles. 1958. *Elementary Seismology*. New York: W. H. Freeman.

Ruiz, S., M. Metois, A. Fuenzalida, J. Ruiz, F. Leyton, R. Grandin, C. Vigny, R. Madariaga, and J. Campos. 2014. "Intense Foreshocks and a Slow Slip Event Preceded the 2014 Iquique Mw 8.1 Earthquake." *Science* 345:1165–69.

Ryan, W. B. F., S. M. Carbotte, J. O. Coplan, S. O'Hara, A. Melkonian, R. Arko, R. A. Weissel, V. Ferrini, A. Goodwillie, F. Nitsche, J. Bonczkowski, and R. Zemsky. 2009. "Global Multi-resolution Topography Synthesis." *Geochemistry, Geophysics, Geosystems* 10, no. 3: Q03014. doi:10.1029/2008GC002332.

Satake, K., Y. Fujii, T. Harada, and Y. Namegaya. 2013. "Time and Space Distribution of Coseismic Slip of the 2011 Tohoku Earthquake as Inferred from Tsunami Waveform Data." *Bulletin of the Seismological Society of America* 103, no. 2B: 1473–92.

Sato, M., M. Fujita, Y. Matsumoto, T. Ishikawa, H. Saito, M. Mochizuki, and A. Asada. 2013. "Interplate Coupling off Northeastern Japan Before the 2011 Tohoku-Oki Earthquake, Inferred from Seafloor Data." *Journal of Geophysical Research* 118:3860–69.

Scholz, C. H. 2002. *The Mechanics of Earthquakes and Faulting.* 2nd ed. Cambridge: Cambridge University Press.

Scholz, C. H., L. R. Sykes, and Y. P. Aggarwal. 1973. "Earthquake Prediction: A Physical Basis." *Science* 181:803–10.

Schurr, B., G. Asch, S. Hainzl, J. Bedford, A. Hoechner, M. Palo, et al. 2014. "Gradual Unlocking of Plate Boundary Controlled Initiation of the 2014 Iquique Earthquake." *Nature* 512:299–312.

Socquet, A., J. P. Valdes, J. Jara, F. Cotton, A. Walpersdorf, N. Cotte, S. Specht, F. Ortega-Calaciati, D. Carrizo, and E. Norabuena. 2017. "An 8 Month Slow Slip Event Triggers Progressive Nucleation of the 2014 Chile Megathrust." *Geophysical Research Letters* 44:4046–53.

Stone, Richard. 2016. "Near Miss at Fukushima Is a Warning for U.S., Panel Says." *Science* 352:1039–40.

Sykes, L. R. 1960. "An Experimental Study of Compressional Velocities in Deep Sea Sediments." BS and MS thesis, MIT, 1960. (This thesis will be deposited along with my other papers in the Rare Book and Manuscript Library at Columbia University.)

——. 1963. "Seismicity of the South Pacific Ocean." *Journal of Geophysical Research* 68:5999–6006.

——. 1964. "Deep-Focus Earthquakes in the New Hebrides Region." *Journal of Geophysical Research* 69:5353–55.

——. 1965. "The Seismicity of the Arctic." *Bulletin of the Seismological Society of America* 55:536–91.

——. 1966. "Seismicity and Deep Structure of Island Arcs." *Journal of Geophysical Research* 71:2981–3006.

——. 1967. "Mechanism of Earthquakes and Nature of Faulting on the Mid-Oceanic Ridges." *Journal of Geophysical Research* 72:2131–53.

——. 1971. "Aftershock Zones of Great Earthquakes, Seismicity Gaps, and Earthquake Prediction for Alaska and the Aleutians." *Journal of Geophysical Research* 76:8021–41.

——. 2017. *Silencing the Bomb: One Scientist's Quest to Halt Nuclear Testing.* New York: Columbia University Press.

Sykes, L. R., J. Armbruster, W.-Y. Kim, and L. Seeber. 2008. "Observations and Tectonic Setting of Historic and Instrumentally-Located Earthquakes in the Greater New York City–Philadelphia Area." *Bulletin of the Seismological Society of America* 98:1696–719.

Sykes, L. R., and J. F. Evernden, 1982. "The Verification of a Comprehensive Nuclear Test Ban." *Scientific American* 247:47–55.

Sykes, L. R., and J. B. Hersey. 1960. "Correlation of Physical Properties of Deep Sea Sediments with Sub-bottom Reflections." Unpublished manuscript. (This manuscript will be deposited along with my other papers in the Rare Book and Manuscript Library at Columbia University.)

Sykes, L. R., and S. Jaumé. 1990. "Seismic Activity on Neighboring Faults as a Long-Term Precursor to Large Earthquakes in the San Francisco Bay Area." *Nature* 348:595–99.

Sykes, L. R., and M. Landisman. 1964. "The Seismicity of East Africa, the Gulf of Aden, and the Arabian and Red Seas." *Bulletin of the Seismological Society of America* 54:1927–40.

Sykes, L. R., M. Landisman, and Y. Sato. 1962. "Mantle Shear Wave Velocities Determined from Oceanic Love and Rayleigh Wave Dispersion." *Journal of Geophysical Research* 67:5257–71.

Sykes, L. R., and W. Menke. 2006. "Repeat Times of Large Earthquakes: Implications for Earthquake Mechanics and Long-Term Prediction." *Bulletin of the Seismological Society of America* 96:1569–96.

Sykes, L. R., and S. Nishenko. 1984. "Probabilities of Occurrence of Large Plate Rupturing Earthquakes for the San Andreas, San Jacinto, and Imperial Faults, California, 1983–2003." *Journal of Geophysical Research* 89:5905–27.

Sykes, L. R., and J. O. Oliver. 1964. "The Propagation of Short-Period Seismic Surface Waves Ccross Oceanic Areas, Part I—Theoretical study, Part II—Analysis of Seismograms." *Bulletin of the Seismological Society of America* 54:1349–415.

Sykes, L. R., E. C. Robertson, and M. Newell. 1961. "Experimental Consolidation of Calcium Carbonate Sediment." In P. E. Cloud Jr., P. D. Blackmon, F. D. Sisler, H. Kramer, J. H. Carpenter, E. C. Robertson, L. R. Sykes, and M. Newell, *Environment of Calcium Carbonate Deposition West of Andros Island Bahamas*, 82–83. U.S. Geological Survey Professional Paper 350. Washington, D.C.: U.S. Geological Survey.

Sykes, L. R., B. E. Shaw, and C. H. Scholz. 1999. "Rethinking Earthquake Prediction," *Pure and Applied Geophysics* 155:207–32.

Tobin, D. G., and L. R. Sykes. 1968. "Seismicity and Tectonics in the Northeast Pacific Ocean." *Journal of Geophysical Research* 73:3821–45.

U.S. Geological Survey (USGS). 1990a. "The Loma Prieta, California, Earthquake: An Anticipated Event." *Science* 247:286–93.

——. 1990b. *The San Andreas Fault System, California*. Professional Paper no. 1515. Washington, D.C.: USGS.

——. 2016. "M 7.0 Scenario Earthquake—Hayward-Rodgers Creek; Hayward N + S." https://earthquake.usgs.gov/contactus/golden/neic.php.

U.S. Geological Survey (USGS) Working Group on California Earthquake Probabilities. 1988. *Probabilities of Large Earthquakes Occurring in California on the San Andreas Fault*. USGS Open-File Report. Washington, D.C.: USGS.

——. 1990. *Probabilities of Large Earthquakes in the San Francisco Bay Region, California*. USGS Circular 1053. Washington, D.C.: USGS.

U.S. National Academies of Science, Engineering, and Medicine. 2014. *Lessons Learned from the Fukushima Nuclear Accident for Improving Safety and Security of U.S. Nuclear Plants: Phase 1*. Washington, D.C.: National Academies Press.

——. 2016. *Lessons Learned from the Fukushima Nuclear Accident for Improving Safety of U.S. Nuclear Plants: Phase 2*. Washington, D.C.: National Academies Press.

U.S. Nuclear Regulatory Commission (NRC). 1973. "Appendix A: Seismic and Geologic Siting Criteria for Nuclear Power Plants." Regulations Title 10, Code of Federal

Regulations, Part 100, "Reactor Site Criteria." https://www.nrc.gov/reading-rm/doc-collections/cfr/part100/part100-appa.html.

——. 2011. *Technical Evaluation by the Office of Nuclear Reactor Regulation Related to Plant Restart After the Occurrence of an Earthquake Exceeding the Level of the Operating Basis and Design Basis Earthquakes.* U.S. NRC Doc. ML11308B406. Washington, D.C.: U.S. NRC, November 11.

——. 2013. *Staff Evaluation and Recommendation for Japan Lessons-Learned Tier 3 Issue on Expedited Transfer of Spent Fuel.* COMSECY-13-0030. Washington, D.C.: U.S. NRC, November 12.

U.S. Nuclear Regulatory Commission (NRC), Atomic Safety and Licensing Appeal Board, M.C. Farrar, Chairman. 1977. "In the Matter of Consolidated Edison Company of New York, INC. and Power Authority of the State of New York (Indian-Point, Units 1, 2, and 3." October 12. ALAB-436.

Vine, F. J. 1966. "Spreading of the Ocean Floor: New Evidence." *Science* 154:1405–15.

Vine, F. J., and D. H. Matthews. 1963. "Magnetic Anomalies Over Oceanic Ridges." *Nature* 199:947–49.

Weingarten, M., S. Ge, J. W. Godt, B. A. Bekins, and J. L. Rubinstein. 2015. "High-Rate Injection Is Associated with the Increase in U.S. Midcontinent Seismicity." *Science* 348:1336–40.

Wilson, J. Tuzon. 1965. "A New Class of Faults and Their Bearing on Continental Drift." *Nature* 207:343–47.

Witherspoon, R. 2011. "Nuclear Plants Face System-Wide Earthquake Safety Review." *Energy Matters*, September 2. www.RogerWitherspoon.com.

Yamanaka, Y., and M. Kikuchi. 2004. "Asperity Map Along the Subduction Zone in Northeastern Japan Inferred from Regional Seismic Data." *Journal of Geophysical Research* 109, no. B7: 1–41.

Yokota, Y., and K. Koketsu. 2015. "A Very Long-Term Transient Event Preceding the 2011 Tohoku Earthquake." *Nature Communications* 6:1–5.

Yue, H., and T. Lay. 2013. "Source Rupture Models for the Mw 9.0 2011 Tohoku Earthquake from Joint Inversions of High-Rate Geodetic and Seismic Data." *Bulletin of the Seismological Society of America* 103:1242–55.

INDEX

Note: Page numbers in italics refer to figures and tables.

ABOUT THE AUTHOR

Dr. Lynn Sykes, along with Walter Pitman of the Lamont-Doherty Earth Observatory of Columbia University and Jason Morgan of Princeton, showed unequivocally that the earth's outermost layers consist of nearly rigid plates that move over the surface of the earth. Referred to as *plate tectonics*, this discovery revolutionized the study of the earth, providing an understanding of the formation of mountain ranges, the drifting of the continents, volcanoes, earthquakes, ocean basins, Mid-Oceanic Ridges, deep-sea trenches, the evolution of climate, and the distribution of natural resources. Sykes's research illustrated the importance of great faults that intersect midocean ridges in accommodating plate motion and on the underthrusting of plates at subduction zones. Dr. Marcia McNutt, a geophysicist and president of the U.S. National Academy of Sciences, called the discovery of plate tectonics "one of the top-ten scientific accomplishments of the second half of the twentieth century." The three scientists were awarded the prestigious Vetlesen Prize in 2000.

For more than fifty-five years, Sykes has been involved in the identification of underground nuclear testing and the long battle to obtain a total ban on nuclear testing. In 1986, the Federation of American Scientists presented Sykes and two colleagues with its Public Service Award for "leadership in applying Seismology to the banning of nuclear tests, creative in utilizing their science, effective in educating their nation, fearless and tenacious in struggles within the bureaucracy."

After graduating with bachelor's and master's degrees in geology from MIT in 1960, Sykes entered Columbia, where he earned his Ph.D. in seismology in 1965. He became a faculty member in 1968 and was named the Higgins Professor of Earth and Environmental Sciences. He taught geophysics, plate tectonics, and environmental hazards. Sykes became a member of the staff of the Lamont-Doherty Earth Observatory of Columbia University in 1965 and remained there until 2005, when he retired as a professor emeritus.

Sykes was a member of the U.S. delegation that traveled to the Soviet Union in 1974 to negotiate the Threshold Test Ban Treaty. He testified before the U.S. Congress numerous times as an expert on nuclear-test verification, a subject with large scientific and public-policy components, and has authored more than 135 scientific articles, including 40 on nuclear testing.

Dr. Sykes is a member of the National Academy of Sciences and the American Academy of Arts and Sciences and a fellow of the Geological Society of America and the American Geophysical Union, which honored him with its Macelwane and Bucher Awards. He also received the Seismological Society of America's prestigious H. F. Reid Medal and the U.S. Geological Survey's John Wesley Powell Award for work on earthquakes in the United States. Although officially retired, he continues his research on earthquakes and nuclear explosions. His first book, *Silencing the Bomb: One Scientist's Quest to Halt Nuclear Testing*, was released by Columbia University Press in the fall of 2017. He was born in 1937 in Pittsburgh, Pennsylvania, and grew up near Washington, D.C. A longtime resident of Palisades, New York, he is married to Kathleen Mahoney Sykes.